Synthesis Lectures on Emerging Engineering Technologies

This series publishes short books on current engineering technologies that are gaining prominence, as well as promising technologies that are being developed, for an audience of researchers, advanced students, engineers and other professionals, and entrepreneurs.

Yasha Yi

From 2D to 3D Photonic Integrated Circuits

Springer

Yasha Yi
University of Michigan
Troy, MI, USA

ISSN 2381-1412   ISSN 2381-1439 (electronic)
Synthesis Lectures on Emerging Engineering Technologies
ISBN 978-3-031-91507-9   ISBN 978-3-031-91508-6 (eBook)
https://doi.org/10.1007/978-3-031-91508-6

© The Editor(s) (if applicable) and The Author(s), under exclusive license to Springer Nature Switzerland AG 2026

This work is subject to copyright. All rights are solely and exclusively licensed by the Publisher, whether the whole or part of the material is concerned, specifically the rights of translation, reprinting, reuse of illustrations, recitation, broadcasting, reproduction on microfilms or in any other physical way, and transmission or information storage and retrieval, electronic adaptation, computer software, or by similar or dissimilar methodology now known or hereafter developed.
The use of general descriptive names, registered names, trademarks, service marks, etc. in this publication does not imply, even in the absence of a specific statement, that such names are exempt from the relevant protective laws and regulations and therefore free for general use.
The publisher, the authors and the editors are safe to assume that the advice and information in this book are believed to be true and accurate at the date of publication. Neither the publisher nor the authors or the editors give a warranty, expressed or implied, with respect to the material contained herein or for any errors or omissions that may have been made. The publisher remains neutral with regard to jurisdictional claims in published maps and institutional affiliations.

This Springer imprint is published by the registered company Springer Nature Switzerland AG
The registered company address is: Gewerbestrasse 11, 6330 Cham, Switzerland

If disposing of this product, please recycle the paper.

Preface

The rapid evolution of photonic integrated circuits (PICs) has played a transformative role in advancing modern technologies, from telecommunications and artificial intelligence to quantum computing and biomedical sensing. Over the past few decades, photonics has emerged as a disruptive force, offering high-speed, low-power, and compact solutions that complement or even surpass electronic circuits in various applications. The transition from two-dimensional (2D) photonic integration to three-dimensional (3D) photonic integration marks a pivotal shift in the way we design, manufacture, and implement photonic devices. As industries seek greater performance, efficiency, and scalability, 3D photonics presents itself as a compelling solution, pushing the boundaries of what is possible in optical technologies.

This book serves as both an introduction and an advanced reference for researchers, engineers, and students working in photonic integration, microelectronics, and semiconductor manufacturing. By merging foundational principles with real-world applications, this work provides a comprehensive exploration of the challenges, breakthroughs, and future directions of 3D photonic integrated circuits. The interdisciplinary nature of photonic integration requires expertise in materials science, nanofabrication, optical physics, and electrical engineering, making this a field that bridges multiple scientific domains. The aim of this book is to not only present cutting-edge advancements in 3D photonics but also to offer insights into the practical implementation and commercial impact of these technologies.

The book begins by laying the foundation of photonic integration, tracing its historical evolution from early optical waveguides to modern silicon photonics and heterogeneous integration techniques. Key topics include:

- Advanced materials for photonic integration (such as silicon, silicon nitride, III-V compounds, and lithium niobate),
- Fabrication techniques (including lithography, bonding, and wafer stacking),
- Thermal management and optical loss reduction strategies, and

- The role of photonics in AI, LiDAR, quantum computing, and high-performance computing.

A central focus of this book is the emergence of 3D photonic integration, which enables:

- Higher integration densities,
- Improved signal processing capabilities,
- Compact and efficient photonic circuits, and
- Heterogeneous system integration for a broader range of applications.

The transition to 3D photonic architectures has also introduced new challenges, including fabrication scalability, alignment tolerances, and thermal dissipation, all of which are explored in depth throughout this book. Additionally, we highlight the latest breakthroughs in photonic packaging, co-integration with CMOS electronics, and the potential of neuromorphic photonic computing.

One of the key motivations behind writing this book is the rapid growth of the global photonics industry and the increasing demand for highly skilled professionals who can contribute to next-generation photonic devices and systems. As the industry advances, collaborations between academia, government research labs, and industry leaders will be crucial in driving innovation. This book aims to bridge the knowledge gap between academic research and industrial applications, providing insights into both theoretical principles and practical implementations.

Writing this book has been a deeply rewarding journey, and I am immensely grateful to the many individuals who have supported me along the way. First and foremost, I would like to express my heartfelt appreciation to my wife, Lilly Tan, whose unwavering encouragement, patience, and belief in my work have been invaluable. Her support has allowed me to dedicate the countless hours necessary to research, analyze, and compile this book. To my daughter, Grace Yi, and my son, Alexander Yi, your curiosity and love for learning continue to inspire me every day. The long nights of writing and editing were made easier knowing that I have a family that believes in my passion for photonics and innovation.

I would also like to extend my gratitude to my colleagues, collaborators, and students, whose stimulating discussions and research contributions have helped shape the content of this book. Many of the ideas presented here are the result of years of collaboration with leading researchers in photonics, optoelectronics, and semiconductor manufacturing, and I deeply appreciate their insights and perspectives.

Finally, I hope this book serves as a valuable resource for researchers, engineers, and students who are eager to explore the next frontiers in 3D photonic integration. As we stand at the crossroads of semiconductor scaling, artificial intelligence, and next-generation computing, photonics will undoubtedly play a key role in shaping the future.

Whether you are a seasoned researcher, an industry professional, or a student beginning your journey in photonics, I hope you find inspiration in these pages and that this book contributes to your understanding, innovation, and continued exploration of photonic technologies.

Thank you for being part of this journey.

Troy, USA
February 2025

Yasha Yi, (Ph.D. MIT)
Fellow of *Optica* (Optical Society of America)
and Professor of Electrical Engineering

Competing Interests The author has no competing interests to declare that are relevant to the content of this manuscript.

Contents

1 Introduction to 2D and 3D Photonics 1
   1.1   Overview of Photonic Integrated Circuits (PICs): Introduction
          to PICs and Their Significance in Modern Technology 1
   1.2   Historical Context: Evolution from Traditional 2D to 3D Photonic
          Integrated Circuits (PICs) and the Transformative Journey 7
   1.3   Importance and Applications: Key Areas Where 3D Photonics is
          Making an Impact 13

2 Fundamentals of Photonic Integrated Circuits 17
   2.1   Basic Principles of Photonics 17
   2.2   Comparison with Electronic Integration: Differences
          and Similarities Between Photonic and Electronic Circuits 24
   2.3   Key Components of Photonic Integrated Circuits (PICs) 27

3 The Evolution from 2D to 3D PICs 33
   3.1   Limitations of 2D Photonic Integrated Circuits (PICs) 33
   3.2   Technological Advancements: Breakthroughs Enabling
          the Transition to 3D Integration 36
   3.3   Case Studies: Examples of Enhanced Performance with 3D PICs
          Compared to Their 2D Counterparts 40

4 Design and Fabrication Techniques 49
   4.1   CMOS-Compatible Processes: Integration of Photonics
          with Mature CMOS Fabrication Techniques 49
   4.2   Additive Manufacturing in Photonics: Techniques Like
          Two-Photon Polymerization for Creating Intricate 3D Structures 54
   4.3   Advanced Lithography and Etching Methods: Detailed Processes
          for Precise Fabrication 58

5	**Thermal Management in 3D PICs**		65
	5.1	Heat Dissipation Challenges: Issues Arising from Increased Packing Density and Hotspots in 3D Photonic Integrated Circuits	65
	5.2	Innovative Cooling Solutions: Advanced Materials and Techniques for Effective Thermal Management in 3D Photonics	72
	5.3	Materials with High Thermal Conductivity: Selection of Materials to Improve Heat Dissipation in 3D Photonics	75
6	**Alignment and Packaging of 3D PICs**		83
	6.1	3D Photonics Precision Alignment Techniques: Methods to Maintain Optical Signal Integrity and Ensure Accurate Positioning	83
	6.2	Packaging Technologies: Strategies for Robust and Efficient Packaging of 3D PICs	87
	6.3	Inter-layer Optical Interconnects in 3D Photonics: Solutions for Vertical Optical Connections with Low Loss and High Misalignment Tolerance	92
7	**Heterogeneous and Hybrid Integration**		99
	7.1	Combining Photonic and Electronic Materials: Benefits and Challenges of Hybrid Integration	99
	7.2	Hybrid Integration Techniques: Methods to Combine Different Material Systems in a Single Device	108
8	**Applications of 3D Photonics**		115
	8.1	Telecommunications and Data Centers: Enhancing Data Transmission and Processing Capabilities Utilizing 3D Photonics	115
	8.2	Solid-State LiDAR and AI Using 3D Photonics: Improving Performance and Miniaturization for Sensing and Artificial Intelligence Applications	120
	8.3	Quantum Computing and Sensing: Role of 3D PICs in Advancing Quantum Technologies	125
9	**Modeling and Simulation Tools**		133
	9.1	Electronic Design Automation (EDA) for Photonics: Adapting Traditional EDA Tools for Photonic Circuits	133
	9.2	Advanced Simulation Techniques: Modeling Interactions Between Various Layers and Components	138
	9.3	Predictive Modeling of 3D Photonic Structures: Ensuring Optimal Performance and Reliability	143

10 Industry Trends and Future Directions ... 149
- 10.1 Emerging Technologies: Latest Advancements and Future Possibilities in 3D Photonics ... 149
- 10.2 Market Trends and Industry Outlook for 3D Photonics: Analysis of Current Market Dynamics and Future Growth Prospects ... 153
- 10.3 Potential Breakthroughs in 3D Photonics: Areas with the Highest Potential for Innovation and Development ... 156

11 Case Studies and Real-World Applications ... 161
- 11.1 Detailed Analysis of Successful Projects in 3D Photonics: Examination of Real-World Implementations and Their Outcomes ... 161
- 11.2 Lessons Learned and Best Practices: Insights from Industry Leaders and Researchers Working in Photonics ... 165
- 11.3 Potential Pitfalls and Solutions: 3D Photonics Common Challenges and Effective Strategies to Overcome Them ... 169

12 Vision for the Future ... 175
- 12.1 Summary of Key Points in 3D Photonics: Recap of the Main Topics Covered in the Book ... 175
- 12.2 Future Impact of 3D Photonics: Predictions for How 3D Photonics Will Shape Future Technologies ... 179
- 12.3 Forward-Looking Statements: Author's Perspective on the Evolving Landscape of 3D Photonics ... 182

Bibliography ... 187

About the Author

Dr. Yasha Yi is currently Fellow of *Optica* (Optical Society of America), a full professor of Electrical Engineering, was the founding EECE Ph.D. program chair and Provost faculty fellow, *University of Michigan*, Dearborn campus, and LNF, University of Michigan, Ann Arbor campus. He received the Ph.D. degree from the *Massachusetts Institute of Technology (MIT)*, Cambridge, MA, USA, and was a post-doctoral associate with the MicroPhotonics Center, Massachusetts Institute of Technology, Cambridge, MA, USA, where he was involved in research on integrated nano-optoelectronic devices and systems. He had extensive research experiences with the *Los Alamos National Laboratory* and the *3M Corporate Research Laboratory*. He is also a professor affiliate with the Microsystems Technology Laboratory at MIT. He has authored more than hundreds of top journal papers, edited four book/book chapters, and holds 33 issued patents (17 U.S. patents and 16 international patent). He has led several government/industry-funded projects, has been at review panel for NSF, DOE and DOD, and has been a reviewer for leading journals. His research interests are integrated/intelligent chips, renewable energy, smart microsensors; virtual reality/augmented reality, solid-state on-chip LiDAR for autonomous driving/UAVs/robotics; telecommunications; and solid-state lighting.

Introduction to 2D and 3D Photonics

- **Overview of Photonic Integrated Circuits (PICs)**: Introduction to PICs and their significance in modern technology.
- **Historical Context**: Evolution from traditional 2D to 3D PICs and the transformative journey.
- **Importance and Applications**: Key areas where 3D photonics is making an impact, such as telecommunications, data centers, and quantum computing.

1.1 Overview of Photonic Integrated Circuits (PICs): Introduction to PICs and Their Significance in Modern Technology

Introduction to Photonic Integrated Circuits

Photonic Integrated Circuits (PICs) represent a significant advancement in the field of photonics, analogous to the role that electronic integrated circuits (ICs) play in electronics. PICs integrate multiple photonic functions on a single chip, utilizing light (photons) instead of electrons to process and transmit information. This capability allows for the development of devices that are faster, more efficient, and capable of handling larger amounts of data than their electronic counterparts.

PICs have found applications in a variety of fields, including telecommunications, data centers, medical diagnostics, and sensing technologies. In telecommunications, PICs enable high-speed data transmission over optical fibers, greatly enhancing the bandwidth and efficiency of communication networks. They play a crucial role in data centers by facilitating faster data processing and transfer, which is essential for handling the ever-increasing data demands of cloud computing and big data applications.

In the medical field, PICs are used in diagnostic devices that require precise and rapid detection of biological markers, enabling early diagnosis and monitoring of diseases. They also find applications in advanced imaging techniques, such as optical coherence tomography (OCT), which is widely used in ophthalmology.

Sensing technologies also benefit from the integration of PICs, as they enable the development of highly sensitive and compact sensors for environmental monitoring, industrial applications, and security systems. The ability of PICs to perform complex optical functions on a compact and integrated platform makes them a key technology for the future of high-speed communication and advanced computing systems.

The development of PICs involves sophisticated fabrication techniques and materials, such as silicon photonics, which leverages the existing infrastructure and knowledge from the semiconductor industry. This synergy allows for the scalable production of PICs and their integration with electronic components, paving the way for hybrid photonic-electronic systems that combine the best of both worlds.

As research and development in photonic integration continue to advance, we can expect to see even more innovative applications and improvements in performance. PICs are set to revolutionize not only telecommunications and data processing but also areas such as quantum computing, where their ability to manipulate and control light at the quantum level holds great promise.

In summary, Photonic Integrated Circuits are at the forefront of technological innovation, offering unparalleled speed, efficiency, and data handling capabilities. Their diverse applications and the potential for future advancements make them an essential technology for the advancement of communication, computing, and sensing technologies.

Historical Context and Development
The concept of integrating photonic components on a single chip can be traced back to the 1960s and 1970s, paralleling the rapid development of electronic integrated circuits (ICs) during the same period. Early efforts focused on simple components like waveguides, modulators, and photodetectors, which laid the groundwork for future advancements in the field of integrated optics. These early PICs were relatively rudimentary compared to today's technology, consisting of just a few basic optical components for low-complexity applications like optical signal modulation or simple optical switching.

Early Developments (1960s–1980s):
The initial breakthroughs in PICs were driven largely by the development of waveguides and early modulator technologies. Waveguides allowed for the controlled propagation of light within the chip, which mirrored the role of metal interconnects in electronic ICs. However, early materials and fabrication processes were limited, with materials like silica and indium phosphide providing the basis for most photonic circuits.

During the 1980s, the field of fiber optics communication began to grow rapidly, which accelerated interest in PICs for telecommunication applications. The potential to transmit vast amounts of data using light opened up new avenues for integrating optical devices onto chips. However, the main challenge at this stage was developing efficient and scalable manufacturing processes that could produce PICs with the same precision and cost efficiency as electronic ICs.

Advances in Materials and Fabrication (1990s–Early 2000s):
Throughout the 1990s and early 2000s, the field of materials science made significant contributions to the advancement of PICs. The development of new semiconductor materials like gallium arsenide (GaAs) and indium phosphide (InP) allowed researchers to build photonic components such as lasers and photodetectors that could be integrated into chips with greater performance and functionality.

At the same time, fabrication techniques were refined, with advances in lithography, etching, and deposition processes enabling the creation of smaller, more efficient photonic structures. These breakthroughs allowed for increasingly complex PIC designs, with multiple waveguides, lasers, and modulators integrated on a single chip.

The Silicon Photonics Revolution (2000s–Present):
One of the most transformative milestones in the evolution of PICs came with the development of the silicon photonics platform in the early 2000s. Silicon photonics leverages the well-established CMOS (complementary metal-oxide-semiconductor) manufacturing techniques that had long been used in the semiconductor industry to produce electronic ICs. This innovation enabled the integration of photonic components with electronics, all on a single chip, using the same high-precision, low-cost manufacturing processes that had been perfected over decades for electronic devices.

The introduction of silicon as an optical material was a major shift. While silicon had long been used in electronics, it was not initially seen as suitable for photonic applications due to its indirect bandgap, which limits its efficiency as a light emitter. However, silicon photonics focused on passive devices like waveguides, modulators, and multiplexers, which did not require active light generation. Lasers could be coupled externally to the silicon chip.

The integration of photonics with mature electronic fabrication techniques allowed for mass production of photonic components at significantly lower costs compared to earlier bespoke processes. It also enabled higher precision, allowing for the fabrication of increasingly smaller and more complex photonic structures. Furthermore, scalability became a key advantage, making it possible to produce PICs with a higher number of integrated components, allowing for more sophisticated applications in data centers, telecommunications, and sensing.

Moreover, this advancement made PIC technology more accessible to industries outside of academia and defense, driving significant interest from commercial sectors such as telecommunications, where bandwidth and speed are critical. Major companies like Intel, IBM, and Cisco have heavily invested in silicon photonics for use in data centers and network infrastructure, where the ability to transmit vast amounts of data optically is essential for managing increasing internet traffic.

The evolution of PICs, particularly with the advent of silicon photonics, has been crucial in enabling modern applications in high-speed data transmission, cloud computing, and optical communication. Additionally, PICs are now finding applications in fields such as sensing and LIDAR, where integrated photonic circuits are used in autonomous vehicle technology and precision sensing. In quantum computing, PICs are enabling new architectures for quantum information processing. In healthcare and biosensing, PICs are being used for lab-on-a-chip technologies that allow for real-time sensing and analysis of biological samples.

The development of silicon photonics has been pivotal in pushing the boundaries of what PICs can achieve. By leveraging existing CMOS technology, the integration of photonics and electronics has led to more powerful, efficient, and scalable solutions that are transforming industries ranging from telecommunications to healthcare.

Key Components of Photonic Integrated Circuits

Photonic Integrated Circuits (PICs) are advanced optical devices that integrate multiple photonic components on a single chip, enabling the manipulation, transmission, and detection of light for various applications. PICs play a crucial role in fields such as telecommunications, data centers, sensing, and quantum computing. Some of the most important components in PICs include waveguides, lasers, modulators, photodetectors, and filters/multiplexers.

Waveguides function similarly to electrical wires but carry light instead of electrical current. They guide light through the chip by confining it within a core material with a higher refractive index than the surrounding cladding. Common materials for waveguides include silicon (for integration with CMOS technology), indium phosphide (preferred for active photonic devices like lasers), and silica. These waveguides are designed to minimize optical losses, ensuring efficient light transmission over long distances on the chip.

Lasers serve as the light sources for photonic circuits, generating coherent light that can be modulated to carry information or perform other optical functions. In PICs, lasers can be either integrated directly onto the chip or coupled externally. Direct integration with materials like indium phosphide enhances performance and reduces coupling losses, whereas external lasers coupled via fiber optics provide flexibility in source selection. Common types of lasers used in PICs include Distributed Feedback (DFB) lasers, Vertical-Cavity Surface-Emitting Lasers (VCSELs), and mode-locked lasers for ultra-fast applications.

Modulators control properties of light such as intensity, phase, or wavelength, enabling the encoding of information onto light signals, similar to electrical modulation in traditional circuits. Electro-optic modulators use the electro-optic effect to change the

1.1 Overview of Photonic Integrated Circuits (PICs): Introduction ...

refractive index of waveguide materials, with silicon-based modulators leveraging the plasma dispersion effect, while lithium niobate provides high-speed modulation. Thermo-optic modulators achieve modulation through heating, offering a simpler but slower approach. These modulators are critical for high-speed data transmission and optical communication applications.

Photodetectors convert light signals back into electrical signals, allowing PICs to interface with electronic systems. This conversion is essential in communication systems, sensors, and applications where optical information must be processed electronically. Common photodetectors in PICs include PIN photodiodes and avalanche photodiodes (APDs), known for their wavelength sensitivity and fast response times. Materials like indium phosphide or gallium arsenide are typically used depending on the targeted wavelength range, such as infrared for telecommunications.

Filters and multiplexers are essential for managing light signals in PICs, enabling the filtering of specific wavelengths or the combination of multiple light signals into a single waveguide. Wavelength-division multiplexers (WDMs) combine multiple signals at different wavelengths into a single waveguide for transmission, while demultiplexers separate them at the receiving end. These components are crucial in optical networks, where transmitting multiple signals over the same physical medium increases bandwidth without requiring additional infrastructure.

The development of these components has significantly enhanced the capabilities of PICs, allowing for advancements in high-speed communication, data processing, and sensing technologies. As integration techniques continue to improve, PICs will further revolutionize industries ranging from telecommunications to healthcare.

PICs integrate a range of photonic components, including waveguides, lasers, modulators, photodetectors, and filters, to perform complex optical functions on a single chip. This integration reduces power consumption, size, and cost while increasing performance, making PICs indispensable in modern photonic applications such as telecommunications, data centers, and sensing.

Significance in Modern Technology
PICs are revolutionizing several key areas of technology due to their unique advantages over traditional electronic circuits.

Telecommunications: The ability of PICs to handle high data rates and operate at low power makes them ideal for telecommunications applications. Optical communication systems based on PICs can transmit data over long distances with minimal loss and at speeds far exceeding those of electronic systems. This capability is essential for meeting the growing demand for high-speed internet and data services.

Data Centers: In data centers, PICs are used to improve the speed and efficiency of data transfer between servers. The integration of optical interconnects based on PICs can significantly reduce power consumption and increase the bandwidth of data center networks, addressing the challenges posed by the exponential growth in data traffic.

Medical Diagnostics and Sensing: PICs are also making an impact in the field of medical diagnostics and sensing. They enable the development of compact and highly sensitive biosensors that can detect specific biomolecules or changes in the environment. These sensors are used in applications ranging from point-of-care diagnostics to environmental monitoring.

Advanced Computing: The integration of photonic circuits with electronic processors is paving the way for new computing paradigms, such as optical computing and neuromorphic computing. These technologies have the potential to overcome the limitations of traditional electronic processors, offering higher processing speeds and lower energy consumption.

Challenges and Future Directions
Despite their numerous advantages, PICs also face several challenges that need to be addressed to fully realize their potential. One of the main challenges is the efficient integration of photonic components with electronic circuits. While silicon photonics has made significant strides in this area, further advancements are needed to improve the performance and scalability of hybrid photonic-electronic systems.

Another challenge is thermal management. PICs, like their electronic counterparts, generate heat during operation. Efficient heat dissipation mechanisms are essential to ensure the reliable performance of PICs, especially as they become more complex and densely packed.

Looking ahead, the future of PICs is promising. Ongoing research and development efforts are focused on improving the efficiency, scalability, and integration capabilities of PICs. Innovations in materials science, such as the development of new photonic materials and nanostructures, are expected to drive the next generation of PICs. Additionally, interdisciplinary collaboration between researchers in photonics, electronics, materials science, and other fields will be crucial in overcoming the challenges and unlocking the full potential of PIC technology.

Photonic Integrated Circuits (PICs) are at the forefront of modern technology, offering unparalleled capabilities in terms of speed, efficiency, and data handling. Their significance spans across various fields, from telecommunications and data centers to medical diagnostics and advanced computing. As the technology continues to evolve, PICs are poised to play a critical role in shaping the future of high-speed communication, data processing, and sensing technologies. Through continued innovation and collaboration, the full potential of PICs will be realized, paving the way for a new era of photonic and electronic integration.

1.2 Historical Context: Evolution from Traditional 2D to 3D Photonic Integrated Circuits (PICs) and the Transformative Journey

Introduction

The field of photonic integrated circuits (PICs) has undergone a significant transformation over the past few decades. Initially rooted in the traditional 2D fabrication processes, PICs have now begun a transformative journey towards intricate 3D configurations. This evolution marks a pivotal shift in how we design, fabricate, and deploy photonic devices, offering unprecedented opportunities for enhancing performance, scalability, and functionality. This chapter delves into the historical context of this evolution, highlighting key milestones, technological advancements, and the transformative impact of transitioning from 2D to 3D PICs (Fig. 1.1).

Early Development of 2D Photonic Integrated Circuits

The journey of PICs began in the 1960s and 1970s, drawing inspiration from the rapid advancements in electronic integrated circuits (ICs). Early PICs were relatively simple, consisting of basic components such as waveguides, modulators, and detectors integrated onto a single substrate. The primary motivation behind these early efforts was to leverage the advantages of photonics—such as high-speed data transmission and low power consumption—within a compact and integrated platform.

One of the pioneering efforts in PIC development was the integration of optical waveguides on a silicon substrate. Silicon, with its well-established fabrication processes from the electronics industry, provided an ideal platform for developing integrated photonic devices. However, these early 2D PICs faced significant challenges in terms of scalability, integration density, and performance limitations due to the planar nature of the designs.

Advancements in 2D PIC Technology

Throughout the 1980s and 1990s, significant advancements were made in the design and fabrication of 2D PICs. Researchers developed more sophisticated components, including high-speed modulators, tunable lasers, and multiplexers, which enabled the integration of more complex photonic functions on a single chip. The use of materials such as indium phosphide (InP) and silicon-on-insulator (SOI) further enhanced the performance and versatility of 2D PICs.

During this period, the telecommunications industry emerged as a major driving force behind PIC development. The demand for high-speed, high-capacity optical communication systems spurred innovations in 2D PIC technology. For instance, the advent of dense wavelength-division multiplexing (DWDM) allowed multiple wavelengths of light to be transmitted simultaneously through a single optical fiber, significantly increasing data transmission capacity.

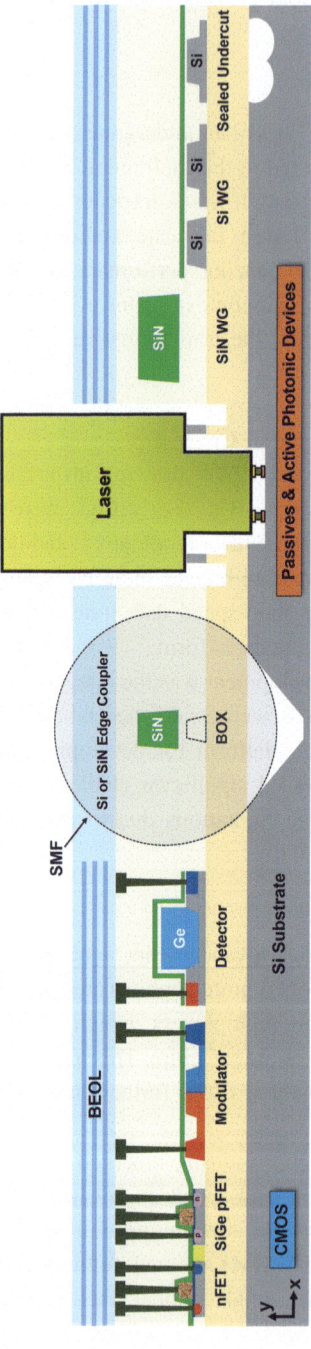

Fig. 1.1 Diagram showing the evolution from 2D to 3D photonic integrated circuits (PICs), illustrating the differences in integration density and structure. Overview of the monolithic SiPh platform which simultaneously integrates laser attach, V-groove-based fiber attach (Si or SiN edge coupler), Si and SiN-based passive and active photonic building blocks as well as CMOS components. By utilizing dual Si thicknesses and dual contact modules, high performance photonic and CMOS devices are enabled and seamlessly integrated on the same silicon-on-insulator (SOI) wafer. The III–V laser source is flip-chip bonded in the SiPh cavity and coupled to a spot-size converter with the assistance of mechanical stops on the SOI substrate along with optical alignment features on the photonic integrated circuit and laser. Drawing not to scale. https://doi.org/10.1109/JSTQE.2023.3238290

Despite these advancements, the planar nature of 2D PICs imposed inherent limitations on their scalability and integration density. The need for longer interconnection paths between components resulted in higher power consumption and reduced operational speeds. These challenges prompted researchers to explore new paradigms that could overcome the limitations of 2D PICs.

Transition to 3D Photonic Integrated Circuits
The transition from 2D to 3D PICs represents a paradigm shift in the field of photonics. The concept of 3D integration involves stacking multiple layers of photonic components vertically, rather than confining them to a single plane. This approach leverages the third dimension to achieve higher integration densities, shorter interconnection paths, and improved overall performance.

One of the key enablers of 3D PICs is the development of through-silicon vias (TSVs), which provide vertical electrical connections between different layers of a stacked chip. TSVs, originally developed for 3D electronic ICs, have been adapted for use in photonic integration, enabling the creation of complex 3D photonic structures. These structures can incorporate multiple layers of waveguides, modulators, detectors, and other photonic components, all interconnected through vertical pathways.

The transition to 3D PICs also necessitated advancements in fabrication techniques. Traditional planar fabrication methods were not sufficient for creating intricate 3D structures. Researchers turned to advanced lithography, etching, and bonding techniques to precisely align and stack multiple layers of photonic components. Innovations in wafer bonding and vertical optical interconnects played a crucial role in achieving the high alignment precision required for 3D PICs (Fig. 1.2).

Advantages of 3D Photonic Integration
The transition from 2D to 3D photonic integration brings numerous benefits, addressing the limitations faced by 2D Photonic Integrated Circuits (PICs) and enabling new levels of performance and functionality:

Enhanced scalability and integration density are among the most significant advantages of 3D Photonic Integrated Circuits (PICs). By utilizing vertical stacking, 3D PICs achieve significantly higher integration densities, allowing for the design of more complex circuits within a smaller footprint compared to their 2D counterparts. This increased density and complexity benefit applications requiring high performance and miniaturization, such as data centers, where 3D PICs improve data throughput while reducing power consumption. In advanced telecommunications systems, greater integration density supports higher data rates and more efficient signal processing. Additionally, solid-state LiDAR systems, used in autonomous vehicles and robotics, rely on compact and high-performance photonic solutions for precision sensing and mapping.

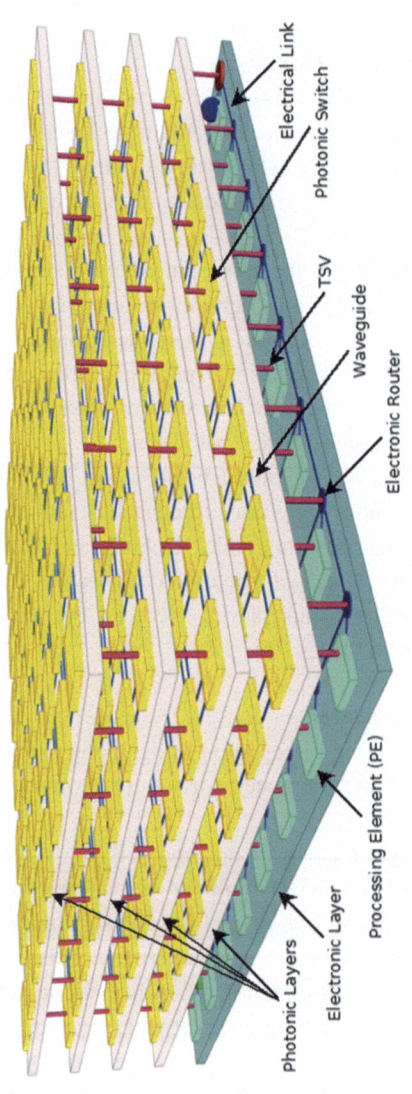

Fig. 1.2 Overview of a 3D PIC with key components labeled (e.g., waveguides, modulators, detectors). PHENIC-II system architecture configured with multiple photonic layers (Photonic Communication Networks (PCNs)). Photonics **2016**, 3, 15; https://doi.org/10.3390/photonics3020015

1.2 Historical Context: Evolution from Traditional 2D to 3D ...

Improved performance and efficiency are also key advantages of 3D PICs. Traditional 2D PICs have interconnects spanning across the chip, leading to increased power consumption and signal delay. In contrast, vertical stacking in 3D PICs significantly shortens these interconnect paths, reducing power consumption by minimizing signal loss and heat generation while also increasing operational speed by reducing latency. This makes 3D PICs ideal for high-speed communication systems and ultra-fast optical computing. Furthermore, stacking multiple functional layers, such as modulators, lasers, and detectors, into a smaller space results in more efficient area utilization, enhancing overall system performance and compactness.

The ability to achieve heterogeneous integration for hybrid systems further enhances the potential of 3D PICs. They enable the integration of different photonic and electronic materials within a single device, such as combining silicon with indium phosphide for photonic applications and gallium arsenide for high-speed electronic devices. This heterogeneous integration allows engineers to leverage the best properties of each material, leading to hybrid systems with enhanced functionality. For instance, photonics can provide high-speed optical data transmission, while electronics handle complex data processing, enabling the combination of lasers, modulators, and photodetectors on a single chip with electronics. This integration improves speed and performance while reducing power consumption.

Despite these advantages, 3D photonic integration presents several technical challenges that must be addressed. Thermal management and heat dissipation are critical concerns, as the high density of components in 3D PICs leads to increased heat generation. The vertical stacking of multiple layers makes heat dissipation more challenging compared to 2D designs. To ensure reliable operation, advanced thermal management techniques are required, including the use of high thermal conductivity materials such as diamond or graphene to efficiently conduct heat away from sensitive components. Additionally, active cooling solutions, such as integrated microfluidic channels, can actively transport heat away. Without proper heat management, elevated temperatures could degrade component performance, particularly in high-power applications like data centers and optical networks.

Precision in alignment and packaging is another major challenge in 3D PICs. The accurate alignment of photonic components across multiple layers is crucial for maintaining optical signal integrity. Misalignment can result in signal loss, crosstalk, and reduced efficiency, especially in applications requiring tight tolerances, such as optical communication systems and sensing devices. Innovations in wafer bonding techniques are essential to ensure that different layers of the chip are bonded with nanometer-level precision. Additionally, vertical optical interconnects must be designed to efficiently transfer light between stacked layers without introducing signal loss. High-precision tools for alignment and packaging will be necessary to overcome these challenges and ensure reliable, scalable production of 3D PICs.

Fabrication complexity also poses a significant hurdle to the widespread adoption of 3D PICs. The multi-layer fabrication process requires significantly more complex techniques compared to traditional 2D photonics. Each layer of the photonic circuit must be fabricated with precision, then aligned and bonded together. Advanced lithography and etching techniques, such as multi-layer patterning and deep etching processes, are required to maintain accuracy across stacked layers. Standardization and the development of process design kits (PDKs) are crucial for streamlining fabrication and making 3D PICs more accessible to a broader range of industries. PDKs provide standardized design rules and component libraries, allowing for more predictable and efficient design, simulation, and fabrication of photonic circuits. However, the complexity of fabricating 3D PICs may initially lead to higher production costs, making them less accessible for certain applications. Over time, as fabrication techniques improve and economies of scale take effect, costs are expected to decrease, enabling broader adoption.

The shift to 3D photonic integration offers transformative advantages, including enhanced scalability, improved performance, and heterogeneous integration, making it a game-changer for applications in data centers, telecommunications, and sensing technologies. However, challenges such as thermal management, alignment precision, and fabrication complexity must be addressed to fully realize the potential of 3D PICs. Ongoing research in materials science, fabrication processes, and design methodologies is critical to overcoming these hurdles and driving the widespread adoption of 3D photonic integration.

Future Directions and Impact
The evolution from 2D to 3D PICs marks a transformative journey in the field of photonics. As researchers continue to push the boundaries of what is possible with 3D integration, the future holds exciting possibilities for advanced photonic systems.

Emerging Applications: The enhanced performance and scalability of 3D PICs open new avenues for applications in telecommunications, data centers, quantum computing, and artificial intelligence. These technologies will drive the next generation of high-speed communication and computing systems.

Interdisciplinary Collaboration: The successful development and deployment of 3D PICs require collaboration across multiple disciplines, including photonics, electronics, materials science, and engineering. Collaborative initiatives and symposiums provide platforms for sharing knowledge, exploring opportunities, and addressing challenges.

Innovation in Design and Simulation: The transition to 3D PICs necessitates new approaches to design and simulation. Advanced simulation techniques and electronic design automation (EDA) tools must be adapted to handle the complexities of 3D photonic circuits, ensuring optimal performance and reliability.

The historical context of the evolution from traditional 2D to 3D PICs highlights the transformative journey that has reshaped the field of photonics. From the early days of simple planar designs to the sophisticated 3D structures of today, PICs have continuously evolved to meet the growing demands of modern technology. As we move forward, the ongoing advancements in 3D photonic integration promise to unlock new levels of performance, scalability, and functionality, paving the way for a future where photonics and electronics work seamlessly together to drive innovation and technological progress.

1.3 Importance and Applications: Key Areas Where 3D Photonics is Making an Impact

Introduction

3D photonics is a rapidly evolving field that holds immense promise for a variety of applications across multiple industries. By leveraging the advantages of three-dimensional integration, photonic integrated circuits (PICs) are pushing the boundaries of what is possible in terms of performance, scalability, and efficiency. This chapter explores the key areas where 3D photonics is making a significant impact, including telecommunications, data centers, and quantum computing. Each of these sectors is benefiting from the unique capabilities of 3D PICs, which are driving innovation and enabling new technological advancements (Fig. 1.3).

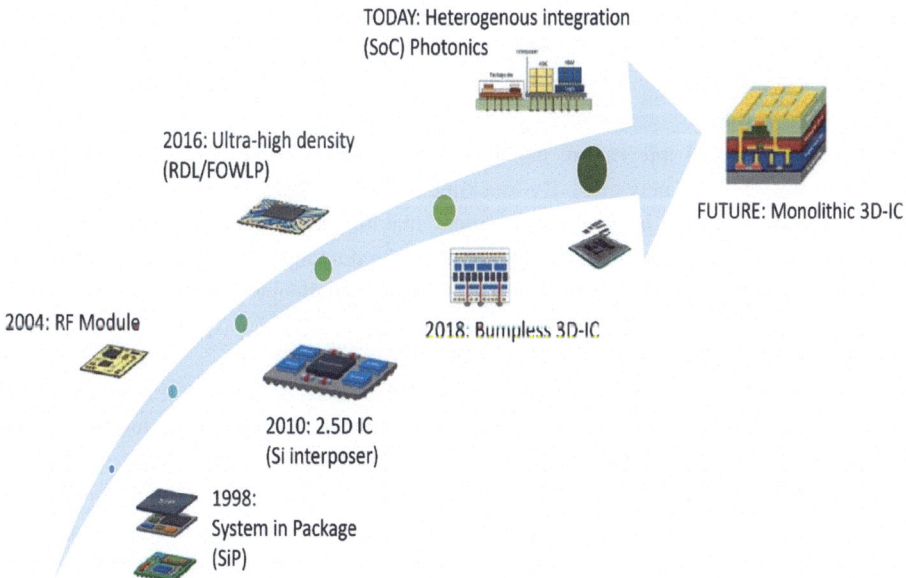

Fig. 1.3 Applications map highlighting where 3D photonics is impacting various industries (e.g., telecommunications, data centers, quantum computing). 3D photonics is impacting various industries (e.g., telecommunications, data centers, quantum computing)

Telecommunications

The telecommunications industry has long been a major driver of advancements in photonic technologies. As the demand for high-speed internet and data services continues to grow, the need for more efficient and capable communication systems becomes increasingly critical. 3D photonics offers several advantages that make it an ideal solution for modern telecommunications networks.

High Data Transmission Rates

One of the primary benefits of 3D photonics in telecommunications is its ability to support high data transmission rates. Traditional electronic circuits struggle to keep up with the bandwidth demands of contemporary communication systems. In contrast, 3D PICs can handle vast amounts of data at speeds that far exceed those of electronic counterparts. This capability is essential for applications such as video streaming, online gaming, and cloud computing, where large volumes of data must be transmitted quickly and reliably.

Reduced Power Consumption

Another significant advantage of 3D photonics is its potential to reduce power consumption in communication systems. Traditional electronic circuits generate considerable heat and require substantial cooling, leading to high energy costs. 3D PICs, on the other hand, operate with greater efficiency, minimizing heat generation and reducing the need for cooling. This efficiency translates into lower operational costs and a smaller environmental footprint, making 3D photonics a more sustainable option for the telecommunications industry.

Enhanced Scalability

The three-dimensional integration of photonic components allows for greater scalability in telecommunications networks. By stacking multiple layers of photonic devices, 3D PICs can achieve higher integration densities, enabling more complex and capable systems. This scalability is particularly important as the demand for data continues to grow, requiring networks to expand and evolve to meet new challenges.

Case Study: Dense Wavelength Division Multiplexing (DWDM)

Dense Wavelength Division Multiplexing (DWDM) is a technology that exemplifies the advantages of 3D photonics in telecommunications. DWDM systems use multiple wavelengths of light to transmit data simultaneously through a single optical fiber, significantly increasing the data-carrying capacity of the network. By integrating DWDM components in a 3D configuration, telecommunications providers can further enhance the performance and efficiency of their networks, supporting higher data rates and greater reliability.

Data Centers

Data centers are another critical area where 3D photonics is making a substantial impact. As the backbone of the digital economy, data centers house vast amounts of data and support a wide range of applications, from cloud computing to artificial intelligence. The integration of 3D photonic technologies offers several key benefits for data centers, addressing some of the most pressing challenges in the industry.

Increased Bandwidth

The ability of 3D PICs to handle high data transmission rates is particularly beneficial for data centers, where large volumes of data must be transferred quickly and efficiently. The increased bandwidth provided by 3D photonics enables faster data processing and reduces latency, improving the overall performance of data center operations.

Energy Efficiency

Energy consumption is a major concern for data centers, which require substantial power to operate and cool their electronic equipment. The efficiency of 3D PICs helps to mitigate this issue by reducing the amount of heat generated and the energy needed for cooling. This energy efficiency translates into lower operational costs and a reduced environmental impact, aligning with the growing emphasis on sustainability in the data center industry.

Compact and Scalable Solutions

The compactness and scalability of 3D photonic components are also significant advantages for data centers. By integrating more functions into a smaller footprint, 3D PICs enable data centers to pack more processing power into limited space. This capability is crucial as data centers continue to expand and evolve, requiring more efficient use of available space to accommodate growing data demands.

Case Study: Optical Interconnects

Optical interconnects based on 3D PICs are a prime example of how this technology is revolutionizing data centers. These interconnects use light to transmit data between servers, reducing latency and increasing data transfer speeds. By integrating optical interconnects in a 3D configuration, data centers can achieve even greater performance and efficiency, supporting the high-speed data processing required for modern applications.

Quantum Computing

Quantum computing represents one of the most exciting and transformative applications of 3D photonics. By leveraging the principles of quantum mechanics, quantum computers have the potential to solve complex problems that are beyond the capabilities of classical computers. 3D photonics plays a crucial role in the development of these advanced computing systems.

Quantum Photonic Circuits
Quantum photonic circuits are the building blocks of quantum computers, enabling the manipulation and measurement of quantum states using light. The three-dimensional integration of these circuits allows for greater complexity and functionality, supporting the development of more powerful quantum computing systems.

High-Density Integration
The ability to integrate multiple layers of photonic components in a 3D configuration is particularly advantageous for quantum computing, where high-density integration is essential. This capability enables the development of compact and scalable quantum processors, which are critical for realizing practical quantum computers.

Improved Performance and Stability
3D photonics also offers improvements in the performance and stability of quantum computing systems. The reduced interconnection lengths and enhanced thermal management provided by 3D integration help to minimize errors and increase the reliability of quantum operations. This stability is crucial for the practical implementation of quantum algorithms and the development of robust quantum computing systems.

Case Study: Quantum Key Distribution (QKD)
Quantum Key Distribution (QKD) is a secure communication method that uses quantum mechanics to encrypt and transmit data. The integration of QKD systems with 3D photonic circuits enhances the performance and security of quantum communication networks. By leveraging the advantages of 3D photonics, QKD systems can achieve higher data transmission rates and greater resistance to eavesdropping, supporting the development of secure quantum communication infrastructure.

Summary
The importance and applications of 3D photonics extend across multiple key areas, including telecommunications, data centers, and quantum computing. The unique capabilities of 3D photonic integrated circuits, such as high data transmission rates, energy efficiency, and scalability, are driving innovation and enabling new technological advancements in these fields. As the technology continues to evolve, the impact of 3D photonics is expected to grow, supporting the development of more efficient, powerful, and sustainable systems across a wide range of applications.

Fundamentals of Photonic Integrated Circuits

- **Basic Principles of Photonics**: Fundamental concepts, including light propagation, waveguides, and photonic crystals.
- **Comparison with Electronic Integration**: Differences and similarities between photonic and electronic circuits.
- **Key Components**: Overview of components like lasers, modulators, detectors, and waveguides used in PICs.

2.1 Basic Principles of Photonics

Photonics, the branch of science and technology focused on the generation, manipulation, and detection of light, has evolved into a pivotal field that underpins a wide range of modern applications. The ability to precisely control light enables innovations that span industries such as telecommunications, healthcare, computing, and sensing. For instance, photonics is the foundation of fiber-optic communication, which allows for the transmission of vast amounts of data over long distances with minimal loss. Similarly, in the realm of medical diagnostics, photonic technologies such as optical coherence tomography (OCT) provide non-invasive, high-resolution imaging of biological tissues.

In this chapter, we will delve into the fundamental principles of photonics, starting with the nature of light propagation and how light behaves when interacting with different materials. We will then explore waveguides, structures that confine and guide light along specific paths, much like electrical wires direct the flow of electricity. Waveguides are essential for integrated photonic circuits, enabling the transmission of optical signals in devices ranging from optical fibers to on-chip photonic systems.

© The Author(s), under exclusive license to Springer Nature Switzerland AG 2026
Y. Yi, *From 2D to 3D Photonic Integrated Circuits*, Synthesis Lectures on Emerging Engineering Technologies, https://doi.org/10.1007/978-3-031-91508-6_2

Next, we will examine photonic crystals, periodic optical nanostructures that can control the flow of light in ways that are impossible with conventional optical materials. Photonic crystals have unique properties that allow them to manipulate light at very small scales, leading to applications in creating more efficient lasers, optical filters, and even in quantum computing.

In addition to photonic crystals, we will explore plasmonics, a field that exploits the interaction between light and free electrons on a metal's surface to create surface plasmon waves. These waves allow light to be confined to extremely small dimensions, far below the diffraction limit of conventional optics. Plasmonics enables applications such as ultrasensitive biosensors, enhanced optical communication, and the development of nanoscale optical devices.

Finally, we will cover metasurfaces, which are engineered surfaces with subwavelength structures designed to manipulate light with unprecedented control. Metasurfaces can alter the phase, amplitude, and polarization of light, enabling the design of ultra-thin optical elements such as flat lenses, beam shapers, and holographic displays. The ability of metasurfaces to compress optical functionality into a 2D structure offers significant advantages for applications in imaging, sensing, and light manipulation at the nanoscale.

By the end of this chapter, you will gain a comprehensive understanding of these core concepts in photonics and their implications for a future where optical technologies play an even more integral role in advancing science, industry, and society (Fig. 2.1).

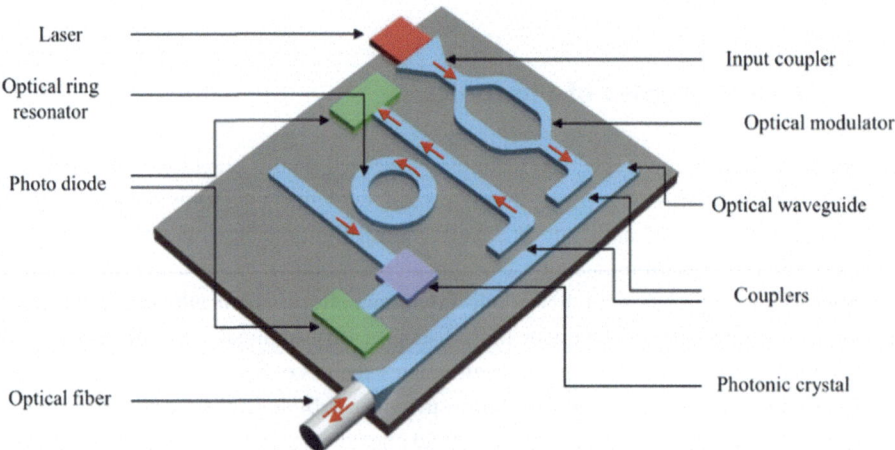

Fig. 2.1 Schematic of light propagation through a waveguide to illustrate basic photonic principles. https://www.ovaga.com/blog/transistor/photonic-integrated-circuit-definition-disadvantage-fabrication-application

2.1 Basic Principles of Photonics

Light Propagation

Light propagation describes how light travels through different media. Understanding this concept is essential for designing and optimizing photonic devices.

Nature of Light

Light exhibits both wave-like and particle-like properties, a duality that is fundamental to quantum mechanics. In photonics, we often treat light as an electromagnetic wave described by Maxwell's equations:

$$\nabla \cdot \mathbf{E} = \rho/\epsilon_0 \quad \nabla \cdot \mathbf{B} = 0 \quad \nabla \times \mathbf{E} = -\frac{\partial \mathbf{B}}{\partial t} \quad \nabla \times \mathbf{B} = \mu_0 \mathbf{J} + \mu_0 \epsilon_0 \frac{\partial \mathbf{E}}{\partial t}$$

where $\mathbf{E}$ is the electric field, $\mathbf{B}$ is the magnetic field, ρ is the charge density, $\mathbf{J}$ is the current density, ϵ_0 is the permittivity of free space, and μ_0 is the permeability of free space.

Reflection and Refraction

When light encounters a boundary between two different media, it undergoes reflection and refraction. Reflection follows the law:

$$\theta_i = \theta_r$$

where θ_i is the angle of incidence, and θ_r is the angle of reflection.

Refraction is described by Snell's Law:

$$n_1 \sin(\theta_i) = n_2 \sin(\theta_t)$$

where n_1 and n_2 are the refractive indices of the respective media, and θ_t is the angle of transmission.

Total Internal Reflection

Total internal reflection occurs when light attempts to move from a medium with a higher refractive index to one with a lower refractive index at an angle greater than the critical angle (θ_c):

$$\theta_c = \sin^{-1}\left(\frac{n_2}{n_1}\right)$$

This principle is crucial in guiding light through optical fibers and waveguides.

Dispersion

Dispersion describes how different wavelengths of light travel at different speeds in a medium, leading to the spreading of a light pulse over time. This is characterized by the material's dispersion relation:

$$\beta(\omega) = n(\omega)\frac{\omega}{c}$$

where β is the propagation constant, ω is the angular frequency, and c is the speed of light in a vacuum.

Waveguides
Waveguides are structures that confine and direct light, similar to how electrical wires guide electricity. They are essential components in photonic integrated circuits (PICs).

Optical Waveguides
Optical waveguides consist of a core material with a higher refractive index than the surrounding cladding, allowing light to be confined within the core by total internal reflection. Common types of optical waveguides include planar waveguides, channel waveguides, and optical fibers.

Planar waveguides are thin films of core material deposited on a substrate, where light is confined in one dimension and propagates within the plane. These waveguides are commonly used in integrated photonic circuits and optical signal processing. Channel waveguides, on the other hand, are formed through etching or deposition processes to create a core channel. In these structures, light is confined in two dimensions and guided along the channel, making them useful for more controlled optical pathways in photonic devices. Optical fibers, widely used in telecommunications, function as cylindrical waveguides that guide light through the fiber core, which is surrounded by cladding with a lower refractive index. These fibers enable long-distance, high-speed optical communication with minimal signal loss, forming the backbone of modern data transmission networks.

Modes in Waveguides
The light in a waveguide can propagate in various modes, each characterized by a specific field distribution. The fundamental mode has the lowest cutoff frequency and the simplest field pattern, making it the most stable and efficient for signal transmission. Higher-order modes exhibit more complex field patterns and require higher frequencies to propagate effectively.

Single-mode waveguides support only the fundamental mode, ensuring high precision and minimal dispersion, making them ideal for applications requiring long-distance, high-speed communication. In contrast, multi-mode waveguides support multiple modes, each traveling at different speeds, which leads to modal dispersion. This dispersion can result in signal distortion over long distances, making multi-mode waveguides more suitable for short-distance communication and lower-speed data transmission applications.

2.1 Basic Principles of Photonics

Waveguide Fabrication
Waveguide fabrication techniques involve several processes to define and construct optical waveguides with precise structural and optical properties. Lithography and etching are commonly used to define waveguide patterns on a substrate, allowing for the precise shaping of optical pathways. Deposition techniques, such as chemical vapor deposition (CVD) and physical vapor deposition (PVD), are employed to deposit core and cladding materials, ensuring the necessary refractive index contrast for guiding light. Another method, diffusion, involves introducing dopants into a substrate to modify its refractive index profile, enabling tailored optical characteristics for specific applications. These fabrication techniques are essential in developing high-performance photonic integrated circuits and optical communication systems.

Photonic Crystals
Photonic crystals are periodic optical structures that affect the propagation of light. They create photonic bandgaps, ranges of wavelengths where light cannot propagate through the structure.

Structure and Properties
Photonic crystals can be one-dimensional (1D), two-dimensional (2D), or three-dimensional (3D), depending on the periodicity of their refractive index variation. One-dimensional photonic crystals, also known as Bragg gratings, consist of alternating layers of materials with different refractive indices. These structures reflect specific wavelengths, making them useful as mirrors or optical filters.

Two-dimensional photonic crystals have a periodic structure in two dimensions, creating a photonic bandgap for in-plane light propagation. These crystals are commonly used in waveguides and resonant cavities, enabling precise control of optical signals within integrated photonic circuits. Three-dimensional photonic crystals exhibit periodicity in all three dimensions, leading to complete photonic bandgaps that prevent light propagation in any direction. This capability allows for advanced control of optical waves, making them essential for developing highly complex photonic devices with novel functionalities.

Plasmonics
Plasmonics is a branch of nanophotonics that deals with the study and manipulation of plasmons, which are collective oscillations of free electron gas density in materials, typically at the interface between a metal and a dielectric. When light interacts with metal nanoparticles or nanostructures, it can excite these plasmons, resulting in unique optical properties such as enhanced electromagnetic fields and strong light confinement at sub-wavelength scales. This phenomenon enables applications in various fields, including sensing, imaging, and information processing. Plasmonic sensors, for instance, can detect minute changes in the refractive index near the metal surface, making them highly sensitive for biochemical sensing. Additionally, plasmonics is integral to the development

of advanced photonic devices, including plasmonic waveguides, which can guide light beyond the diffraction limit, and plasmonic nanolasers, which have the potential to revolutionize on-chip optical communications. Overall, the field of plasmonics bridges the gap between photonics and nanotechnology, paving the way for innovative technologies with unprecedented capabilities.

Metasurface
Metasurfaces are ultra-thin, two-dimensional materials engineered to control light at subwavelength scales, offering unprecedented control over the phase, amplitude, and polarization of light. Unlike traditional optical components that rely on gradual changes in refractive index to bend light, metasurfaces use arrays of nano-antennas or meta-atoms, each smaller than the wavelength of light, to locally modulate the properties of incoming waves. This precise manipulation allows for the creation of flat optical devices with functionalities such as lenses, holograms, and beam shapers, which are significantly thinner and lighter than their conventional counterparts. Metasurfaces have applications in a wide range of fields, from improving the efficiency of solar cells by enhancing light absorption to enabling compact and lightweight components for augmented reality devices. Additionally, they hold promise for advancements in imaging systems, such as creating high-resolution, flat optics for cameras and telescopes, and for developing novel light-based communication technologies. The versatility and compactness of metasurfaces make them a transformative technology in the field of photonics, opening up new possibilities for integrating complex optical functionalities into smaller and more efficient devices.

Metasurfaces are revolutionizing various fields with their ability to control light with subwavelength precision. One of the most significant applications is in flat optics, where metasurfaces enable the creation of ultra-thin, flat lenses and optical components that can replace traditional bulky lenses. These flat lenses are used in cameras, smartphones, and other imaging systems to reduce size and weight while maintaining or even enhancing performance.

In holography, metasurfaces can produce high-resolution, dynamic holograms, making them valuable for applications in augmented reality (AR) and virtual reality (VR). These holographic displays provide more realistic and immersive visual experiences. Additionally, metasurfaces allow for precise beam shaping and steering, which is crucial in LiDAR (Light Detection and Ranging) systems used in autonomous vehicles, as well as in free-space optical communications.

Metasurfaces also play a key role in polarization control by manipulating the polarization state of light. This functionality is essential for advanced microscopy techniques, optical communication systems, and the development of polarization-sensitive devices. Optical filtering is another important application, where metasurfaces act as optical filters with tailored transmission and reflection properties, making them useful in spectroscopy, sensors, and imaging systems for selective wavelength filtering.

2.1 Basic Principles of Photonics

Wavefront control is another critical function of metasurfaces, as they can correct optical aberrations and improve image quality in telescopes, microscopes, and other imaging devices. In nonlinear optics, metasurfaces enhance nonlinear optical processes such as frequency doubling, which are essential in laser technology and in generating new wavelengths of light for various scientific and industrial applications.

In sensing applications, metasurfaces enhance sensor sensitivity by concentrating light into small volumes and improving light-matter interactions. This makes them suitable for chemical and biological sensing, environmental monitoring, and health diagnostics. They are also being explored in quantum optics for manipulating quantum states of light and enhancing light-matter interactions at the quantum level, with potential applications in quantum computing, secure communications, and advanced quantum sensing.

Metasurfaces have significant implications for solar energy, improving the efficiency of solar cells by enhancing light absorption and directing more light into the active layers of the cells, thereby increasing overall power conversion efficiency. Additionally, in antennas and radio frequency (RF) devices, metasurfaces enable the creation of high-performance, compact antennas with improved directional control, which are widely used in wireless communication, satellite communication, and radar systems.

In biomedical imaging and therapy, metasurfaces improve imaging techniques such as optical coherence tomography (OCT) and enable new therapies by precisely focusing light on specific tissues. This has significant implications for medical diagnostics and treatment.

Metasurfaces offer a transformative approach to manipulating light, leading to advancements across a wide array of technological fields, from consumer electronics to advanced scientific research. Their versatility and precision continue to drive innovation, making them an essential component in the future of optics and photonics.

Applications

Photonic crystals have numerous applications across various fields. Two-dimensional photonic crystals can function as waveguides, allowing light to be guided around sharp bends with minimal loss. They are also used in resonant cavities, where light is trapped in a small volume to enhance light-matter interactions, making them valuable in laser and sensor applications. Additionally, one-dimensional photonic crystals serve as optical filters, selectively filtering specific wavelengths, which is particularly useful in telecommunications and spectroscopy.

The fabrication of photonic crystals involves advanced techniques to achieve precise nanoscale structures. Electron beam lithography is commonly used to create highly accurate patterns at the nanoscale, enabling precise control over the crystal's optical properties. Nanoimprint lithography provides a high-throughput method for patterning nanoscale features, making it more suitable for large-scale manufacturing. Another approach, self-assembly, leverages the natural tendency of materials to form periodic structures at the nanoscale, offering a cost-effective method for fabricating photonic crystals.

Understanding the basic principles of photonics is crucial for designing and optimizing photonic integrated circuits. Concepts such as light propagation, waveguides, and photonic crystals form the foundation of PIC operation. As technology advances, the ability to manipulate light at ever-smaller scales will continue to drive new applications and innovations in telecommunications, computing, and beyond.

2.2 Comparison with Electronic Integration: Differences and Similarities Between Photonic and Electronic Circuits

Photonic and electronic circuits are at the heart of modern technology, driving advancements in communications, computing, and various other fields. While both types of circuits serve the purpose of processing and transmitting information, they operate on fundamentally different principles and exhibit unique characteristics. This chapter explores the differences and similarities between photonic and electronic circuits, providing a comprehensive understanding of their roles and how they complement each other (Fig. 2.2).

Fundamental Principles

Photonic circuits use light (photons) to perform operations, leveraging the properties of electromagnetic waves in the optical spectrum. These circuits are designed to manipulate light in various ways, including guiding, modulating, switching, and detecting it. Light propagates through waveguides or optical fibers, which confine and direct the light using principles such as total internal reflection. Photonic circuits operate at wavelengths typically in the infrared (IR) or visible spectrum, ranging from 700 to 1600 nm for most practical applications.

Electronic circuits use electrons to carry and process information, relying on the movement of electrical charges through conductive materials, guided by electric fields. In electronic circuits, electrical currents flow through conductors and semiconductors, driven

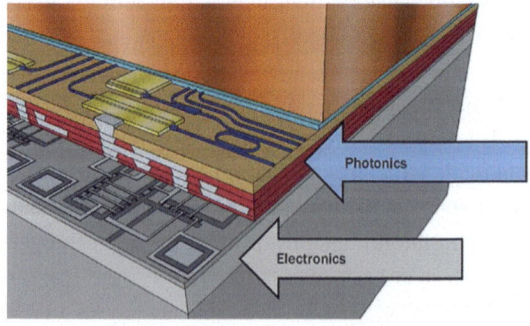

Fig. 2.2 Diagram comparing the structures of photonic and electronic circuits to show their similarities and differences. https://photonicsreport.com/blog/the-fascinating-relationship-between-photonics-and-electronics/

by voltage differences. These circuits typically operate at frequencies up to several gigahertz (GHz) in radio-frequency (RF) applications, but much lower frequencies for most digital circuits.

One of the key differences between photonic and electronic circuits is speed and bandwidth. Photonic circuits can handle extremely high data rates and bandwidths, with optical signals modulating at frequencies in the terahertz (THz) range. This allows for data transmission rates that far exceed those of electronic circuits, making photonics ideal for high-speed data communication and large-scale data transfer applications. Electronic circuits, while generally slower, are well-suited for high-speed processing tasks within the GHz range. However, they face limitations due to resistive losses and capacitance, which slow down signal propagation and limit bandwidth.

Losses and signal integrity also differ between the two technologies. Optical signals in photonic circuits experience minimal propagation losses over long distances compared to electronic signals. This advantage is particularly significant in fiber-optic communications, where signals can travel thousands of kilometers with low attenuation. However, photonic circuits are susceptible to issues like scattering, absorption, and coupling losses. In contrast, electrical signals in electronic circuits suffer from resistive heating, dielectric losses, and electromagnetic interference (EMI), leading to signal degradation over distance and requiring the use of repeaters and amplifiers for long-distance communication.

Power consumption is another distinguishing factor. Photonic circuits generally consume less power for data transmission due to lower resistive losses. However, the power consumption of light sources (lasers) and detectors, as well as the need for precise alignment and coupling, can offset these savings. Electronic circuits, on the other hand, often consume more power due to resistive losses and the need to drive capacitive loads. Power efficiency is a critical factor in electronic circuit design, particularly for battery-powered and portable devices.

Heat generation is a significant challenge for electronic circuits due to resistive losses in conductors and switching losses in transistors. Effective cooling solutions, such as heat sinks and thermal interface materials, are essential to maintain performance and reliability. Photonic circuits, in comparison, generate less heat because light propagation through optical fibers or waveguides does not involve resistive losses. However, thermal management is still crucial in densely packed photonic devices, where heat generated by light sources and modulators needs to be dissipated.

Integration density presents another difference between the two technologies. Photonic components are generally larger than electronic components due to the wavelength of light, which limits the integration density of photonic circuits. Advances in nanophotonics are addressing this challenge, but photonic circuits still lag behind electronic circuits in component density. Electronic circuits benefit from the miniaturization of transistors, with modern integrated circuits (ICs) containing billions of transistors on a single chip. This high integration density enables complex and powerful electronic devices in compact form factors.

Despite these differences, photonic and electronic circuits share several similarities. Both leverage similar fabrication technologies, particularly when photonic circuits are built on silicon platforms. Techniques such as photolithography, etching, deposition, and doping are common in the manufacturing processes of both photonic and electronic components. This compatibility has led to the development of silicon photonics, which integrates photonic components with electronic circuitry on the same chip.

Both types of circuits utilize various modulation techniques to encode information onto carrier waves. In photonics, modulation methods include intensity modulation, phase modulation, and frequency modulation of light. Similarly, electronic circuits use amplitude modulation (AM), frequency modulation (FM), and phase modulation (PM) to encode signals. Signal processing techniques such as amplification, filtering, and switching are also common to both domains.

Hybrid integration is an emerging trend, where photonic and electronic components are combined on a single platform. This approach leverages the high-speed, high-bandwidth capabilities of photonics with the mature processing and control capabilities of electronics. Hybrid systems are finding applications in data centers, telecommunications, and advanced computing, offering enhanced performance and functionality. Additionally, both photonic and electronic circuits aim for scalability in production, with advances in fabrication technologies and materials science enabling large-scale manufacturing. Techniques such as wafer-scale integration and automated assembly processes help reduce costs and improve performance.

Applications of photonic and electronic circuits span multiple industries. In telecommunications, photonic circuits enable high-speed, long-distance data transmission through fiber-optic networks, with technologies like Dense Wavelength Division Multiplexing (DWDM) and coherent optical communication relying on photonics. Electronic circuits, meanwhile, handle signal processing, switching, and routing in telecommunications networks, performing essential tasks such as encoding and decoding data, error correction, and network management.

In data centers, photonic circuits facilitate high-speed interconnects between servers and storage devices, reducing latency and increasing data transfer rates. Electronic circuits, on the other hand, manage data processing, storage, and control functions, providing computational power and memory for handling large volumes of data. In computing and artificial intelligence (AI), photonic circuits are being explored for optical computing and neural network processing, offering potential advantages in speed and energy efficiency. Electronic circuits continue to dominate computing and AI applications, with semiconductor advancements driving improvements in processing power, efficiency, and scalability.

Photonic circuits are also widely used in sensing and imaging applications, including medical diagnostics, environmental monitoring, and industrial inspection. They enable high-resolution, high-sensitivity measurements. Electronic circuits play a crucial role in

2.3 Key Components of Photonic Integrated Circuits (PICs)

signal processing and data acquisition, converting analog signals from sensors into digital data for analysis and interpretation.

While photonic and electronic circuits operate on different physical principles, they share many similarities in fabrication technologies, signal processing techniques, and applications. The unique advantages of each technology make them complementary, with photonics offering high-speed, high-bandwidth capabilities and electronics providing mature, dense integration and processing power. The ongoing trend towards hybrid integration promises to unlock new possibilities, leveraging the strengths of both photonic and electronic circuits to drive innovation and technological progress across various fields. Understanding the differences and similarities between these two types of circuits is crucial for designing and optimizing next-generation systems that can meet the growing demands of our digital world.

2.3 Key Components of Photonic Integrated Circuits (PICs)

Photonic Integrated Circuits (PICs) represent a significant technological advancement, integrating multiple photonic functions on a single chip. The core components of PICs—lasers, modulators, detectors, and waveguides—enable a wide range of applications in telecommunications, data centers, medical diagnostics, and beyond. This chapter provides an overview of these key components, exploring their functions, types, and roles within PICs (Fig. 2.3).

Lasers

Lasers are the primary light sources in PICs, generating coherent light through the process of stimulated emission. The ability to produce a highly focused, monochromatic beam makes lasers indispensable in various applications, from data transmission to sensing.

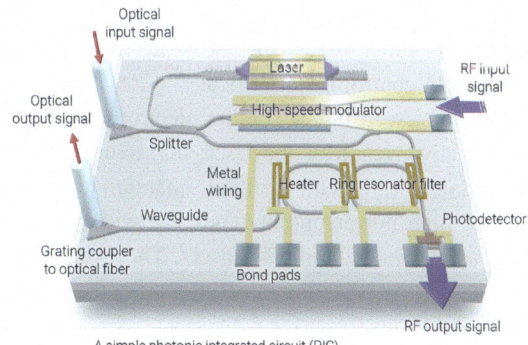

Fig. 2.3 Illustration of key photonic components (lasers, modulators, detectors) within a PIC. https://www.optica-opn.org/home/articles/volume_35/march_2024/features/programmable_photonics/

A simple photonic integrated circuit (PIC).

Types of Lasers
Semiconductor lasers are the most widely used light sources in PICs due to their compact size, high efficiency, and compatibility with semiconductor fabrication processes. These lasers are essential for applications in optical communication, sensing, and data transmission. The two most common types of semiconductor lasers used in PICs are Distributed Feedback (DFB) lasers and Vertical Cavity Surface Emitting Lasers (VCSELs), though other types cater to specific performance needs.

Distributed Feedback (DFB) lasers incorporate a grating structure directly within the semiconductor material, which acts as a feedback mechanism for the laser cavity. The grating selectively reflects certain wavelengths, ensuring that the laser operates at a single wavelength. DFB lasers are known for their extremely narrow linewidth and excellent wavelength stability, making them ideal for long-distance communication in fiber optic networks. These lasers are commonly used in high-speed data communication, telecommunications, and wavelength-division multiplexing (WDM) systems. They offer high stability, low noise, and long operational lifetimes, making them suitable for both terrestrial and submarine optical communication systems.

Vertical Cavity Surface Emitting Lasers (VCSELs) emit light perpendicular to the surface of the semiconductor chip, in contrast to edge-emitting lasers. Their structure consists of a vertically oriented cavity with mirrors that guide the light emission. VCSELs are known for their high power efficiency, low cost, and ability to be mass-produced. They are widely used in short-distance communication systems, such as data centers, optical interconnects, and consumer electronics like 3D sensing in smartphones. VCSEL arrays are easily integrated into high-density PICs, making them suitable for applications requiring parallel processing of multiple optical signals, such as optical switches in data centers.

Fabry-Pérot (FP) lasers use two parallel mirrors to form a resonant cavity, with laser light bouncing back and forth between them. The gain medium amplifies the light to generate coherent output. Unlike DFB lasers, FP lasers often operate over a broader spectrum of wavelengths due to the absence of a built-in grating structure. They are used in less wavelength-sensitive applications, such as short-reach data communication or as pumping sources for other lasers. FP lasers are cost-effective and commonly used where precise wavelength control is not critical.

Quantum well lasers confine electrons and holes within very thin layers of material in the active region of the laser, improving the efficiency of the lasing process. Due to the quantum confinement effect, quantum well lasers offer enhanced efficiency and can achieve high power output. They are widely used in telecommunication systems, fiber-optic networks, and optical sensors. Their versatility allows them to be tuned for different wavelengths by adjusting material properties, enabling diverse applications from long-haul communications to sensor networks.

Tunable lasers allow for dynamic selection of the output wavelength, making them versatile in multi-wavelength applications. The tuning can be achieved through external control of the laser cavity length or through active material manipulation. These lasers are especially valuable in optical networks that require flexibility in assigning wavelengths for different data channels. They are commonly used in WDM systems, which carry multiple data channels over a single optical fiber, greatly increasing network capacity. Tunable lasers are crucial in reconfigurable optical add-drop multiplexers (ROADMs), enabling the dynamic switching of wavelengths in optical networks.

Mode-locked lasers generate ultra-short pulses of light by locking the phases of different modes of the laser cavity, producing a train of pulses with very short durations, typically in the femtosecond range. These lasers are capable of generating pulses at extremely high speeds, making them essential for time-sensitive applications. They are used in optical time-domain reflectometry, ultra-fast data processing, and spectroscopy. Mode-locked lasers are also crucial for developing optical clocks and frequency combs, which have applications in precision measurements and quantum computing.

Semiconductor lasers are the backbone of PICs, enabling efficient light generation for various applications. The selection of a particular type of laser—whether DFB, VCSEL, or another—depends on the specific requirements of the application, such as wavelength precision, power output, and distance. As PIC technology continues to evolve, the integration of these lasers into ever more complex photonic circuits is expected to drive further advancements in telecommunications, data processing, and sensing technologies. Other types of lasers, such as solid-state lasers and fiber lasers, are used in conjunction with PICs for high-power applications but are not typically integrated directly due to their larger size.

Functions of Lasers in PICs
Lasers serve as the primary light sources for transmitting data through optical fibers. The coherent light produced by lasers can be modulated to encode information. They are also used in various sensing applications, providing the light necessary for detecting environmental changes or biological samples. In spectroscopy, lasers provide the precise wavelengths required for analyzing materials based on their spectral characteristics.

Modulators
Modulators control the properties of light, such as its intensity, phase, or wavelength, allowing information to be encoded onto the light signal. They are essential for transmitting data in optical communication systems. Electro-optic modulators use materials whose refractive index changes with applied voltage, such as lithium niobate and silicon. Mach-Zehnder Interferometers (MZIs) split the light into two paths, modulate one path, and then recombine the light, causing interference that encodes the signal. Acousto-optic modulators use sound waves to modulate light, altering its properties through the interaction between acoustic and light waves. These modulators offer high-speed modulation and precise control. Thermo-optic modulators change the refractive index through temperature variations, providing a simpler but slower modulation method.

Functions of Modulators in PICs
Modulators encode data onto light signals for transmission over optical networks. They manipulate light signals in optical signal processing for functions such as filtering and switching. Additionally, they control the phase of light, which is essential in applications like interferometry and coherent optical communication.

Detectors
Photodetectors convert light signals back into electrical signals, enabling the interface between optical and electronic systems. They are crucial for the reception and interpretation of data transmitted through optical networks. PIN photodiodes are widely used in telecommunications due to their high speed and sensitivity. Avalanche photodiodes (APDs) provide internal gain through avalanche multiplication, enhancing sensitivity for detecting weak light signals, making them ideal for long-distance communication. Phototransistors combine a photodiode and a transistor, offering built-in amplification of the detected signal for higher sensitivity applications.

Functions of Detectors in PICs
Detectors receive light signals transmitted through optical fibers and convert them into electrical signals for further processing. In imaging applications, photodetectors capture light from objects, converting it into electrical signals that can be processed into images. They are also used in sensing applications, converting light changes into electrical signals that indicate environmental changes or the presence of specific substances.

Waveguides
Waveguides confine and direct light within PICs, analogous to how electrical wires guide electricity. They are fundamental for routing light signals on a chip. Silicon waveguides offer high refractive index contrast, enabling tight confinement of light and dense integration on a chip, making them widely used in silicon photonics. Silicon nitride waveguides provide lower losses for certain wavelength ranges, making them suitable for longer-distance on-chip communication. Polymer waveguides are flexible and low-cost, used in applications where performance requirements are less stringent.

Functions of Waveguides in PICs
Waveguides direct light signals from one component to another within the PIC, ensuring efficient transmission with minimal loss. They control the mode of light propagation, which is crucial for maintaining signal integrity and minimizing cross-talk between channels. They also serve as optical interconnects, linking different parts of the PIC and enabling complex photonic circuits.

Conclusion
The key components of Photonic Integrated Circuits—lasers, modulators, detectors, and waveguides—each play a crucial role in enabling the diverse functionalities of PICs.

2.3 Key Components of Photonic Integrated Circuits (PICs)

Lasers provide the light source, modulators encode information onto the light, detectors convert light back into electrical signals, and waveguides direct light within the circuit. Together, these components allow PICs to perform a wide range of tasks, from high-speed data transmission to precise sensing and imaging. Understanding the functions and types of these components is essential for designing and optimizing photonic systems that meet the demands of modern technology.

The Evolution from 2D to 3D PICs

3

- **Limitations of 2D PICs**: Challenges related to integration density, interconnection length, and performance.
- **Technological Advancements**: Breakthroughs enabling the transition to 3D integration.
- **Case Studies**: Examples of enhanced performance with 3D PICs compared to their 2D counterparts.

3.1 Limitations of 2D Photonic Integrated Circuits (PICs)

Photonic Integrated Circuits (PICs) are pivotal in advancing telecommunications, data centers, and many other high-tech industries. However, traditional 2D PICs face several significant challenges that limit their performance and scalability. This chapter explores the limitations of 2D PICs, focusing on issues related to integration density, interconnection length, and overall performance (Fig. 3.1).

Integration density refers to the number of components that can be integrated on a single chip. Higher integration density allows for more complex and capable circuits. However, 2D PICs face several barriers to achieving high integration density. The physical size of photonic components is directly related to the wavelength of light they manipulate. For visible and near-infrared wavelengths, the components tend to be relatively large compared to electronic components, limiting the number of photonic elements that can be integrated on a single 2D plane. Adequate spacing between components is necessary to prevent crosstalk and interference. In 2D PICs, maintaining this spacing while increasing the number of components leads to larger chip sizes and limits integration density.

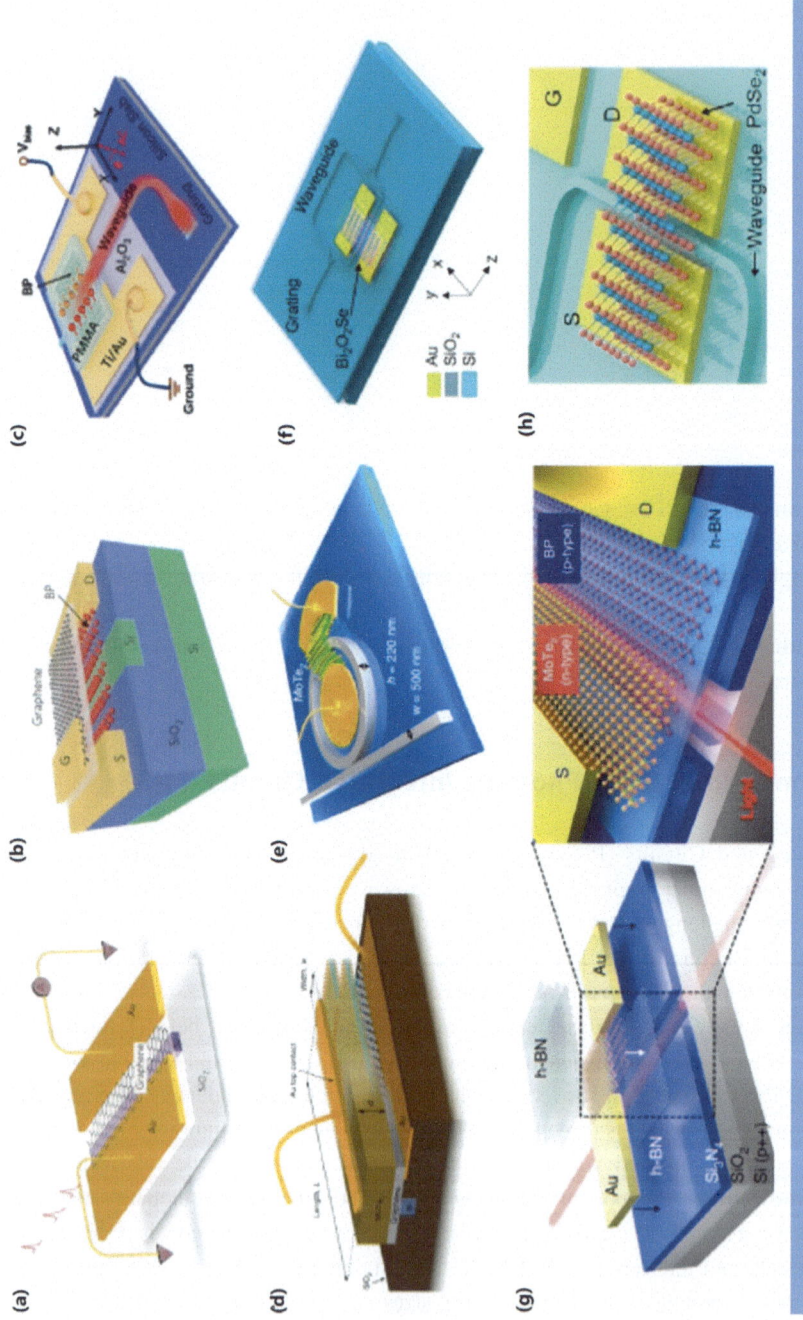

Fig. 3.1 Illustration of limitations in 2D PICs, such as longer interconnections and reduced integration density. https://www.sciencedirect.com/science/article/pii/S2352847822001526

3.1 Limitations of 2D Photonic Integrated Circuits (PICs)

As the number of components on a chip increases, so does the heat generated. Effective thermal management becomes challenging, especially in densely packed 2D PICs. Excessive heat can lead to thermal crosstalk, where the heat from one component affects the performance of adjacent components, degrading overall circuit performance. The materials used in PICs, such as silicon, have specific thermal properties that limit how closely components can be packed together without exceeding thermal limits, further constraining integration density. High integration density also requires extremely precise alignment of photonic components during fabrication. Achieving the necessary precision in a 2D plane is complex and can lead to higher manufacturing costs and lower yields. Variations in the fabrication process can lead to inconsistencies in component performance. In densely integrated 2D PICs, these variations can have significant impacts, making it difficult to achieve uniform performance across all components.

Interconnection length refers to the distance between components on a chip. In 2D PICs, longer interconnections can lead to several performance issues. As light travels through waveguides, it experiences losses due to absorption, scattering, and bending. Longer interconnections mean that signals must travel further, increasing total propagation loss and reducing overall circuit efficiency. Coupling light between different components or waveguides also introduces losses, and longer interconnections typically require more coupling points, further degrading signal integrity. The speed of light in a waveguide is finite, and longer interconnection lengths introduce delays in signal transmission. In high-speed communication systems, even small delays can be significant, affecting the timing and synchronization of signals across the circuit. Longer interconnections can also limit the bandwidth of the signals they carry. Dispersion and other waveguide properties can cause signal degradation over long distances, reducing the effective bandwidth and performance of the PIC.

In densely packed 2D PICs, closely spaced waveguides can cause electromagnetic interference (EMI) between adjacent channels, degrading signal quality and limiting overall performance. Different modes of light propagation within waveguides can couple together, causing interference. In longer interconnections, the probability of mode coupling increases, leading to further performance degradation. Performance in PICs encompasses various factors, including speed, efficiency, and signal quality. 2D PICs face several inherent limitations that affect their overall performance. The speed at which light signals can be modulated is a critical performance metric. In 2D PICs, the physical size and spacing of components can limit the maximum achievable modulation speed. Electrical connections to modulators also introduce resistance and capacitance, which can slow down signal processing. High data throughput requires high-speed modulation and efficient signal routing. The limitations in integration density and interconnection length directly impact the maximum data throughput of 2D PICs.

Generating and maintaining sufficient optical power over long interconnections in 2D PICs requires significant energy. Lasers and amplifiers must compensate for propagation and coupling losses, increasing overall power consumption. Inefficient thermal management leads to higher power consumption and can negatively affect performance and circuit lifespan. The ability to integrate more components and functionalities onto a single chip is essential for scaling PICs. The physical and thermal constraints of 2D integration make it difficult to scale beyond a certain point, limiting the complexity and capabilities of 2D PICs. Scaling up the manufacturing process while maintaining high yields and consistent performance is challenging for 2D PICs. Process variations and alignment issues become more pronounced as the scale increases, affecting the scalability of production.

To address these limitations, researchers and engineers are exploring several approaches, including the transition to 3D PICs, advanced materials, and novel fabrication techniques. Vertical integration in 3D PICs stacks multiple layers of photonic components, significantly increasing integration density without expanding the chip's footprint. This approach reduces interconnection lengths and associated losses. 3D integration also allows for better distribution of heat-generating components, improving thermal management and reducing thermal crosstalk. Using materials with lower optical losses for waveguides and other components can reduce propagation and coupling losses, enhancing overall performance. Materials with high thermal conductivity can improve heat dissipation, allowing for higher integration densities and better performance in both 2D and 3D PICs.

Advanced lithography and etching techniques can achieve higher precision and alignment accuracy, reducing variability and improving integration density. Additive manufacturing techniques like two-photon polymerization enable the fabrication of complex 3D structures on a chip, paving the way for more intricate and capable PICs. While 2D Photonic Integrated Circuits have driven significant advancements in various fields, they face inherent limitations related to integration density, interconnection length, and performance. Addressing these challenges requires innovative approaches, including transitioning to 3D PICs, developing advanced materials, and employing novel fabrication techniques. By overcoming these limitations, PICs can achieve higher performance, greater scalability, and broader application potential, driving the next wave of technological innovation.

3.2 Technological Advancements: Breakthroughs Enabling the Transition to 3D Integration

The evolution from 2D to 3D integration in Photonic Integrated Circuits (PICs) represents a significant leap in technology, offering solutions to many of the limitations inherent in 2D systems. This chapter explores the key technological advancements that have enabled this transition, focusing on breakthroughs in fabrication techniques, materials, and design methodologies (Fig. 3.2).

3.2 Technological Advancements: Breakthroughs Enabling ...

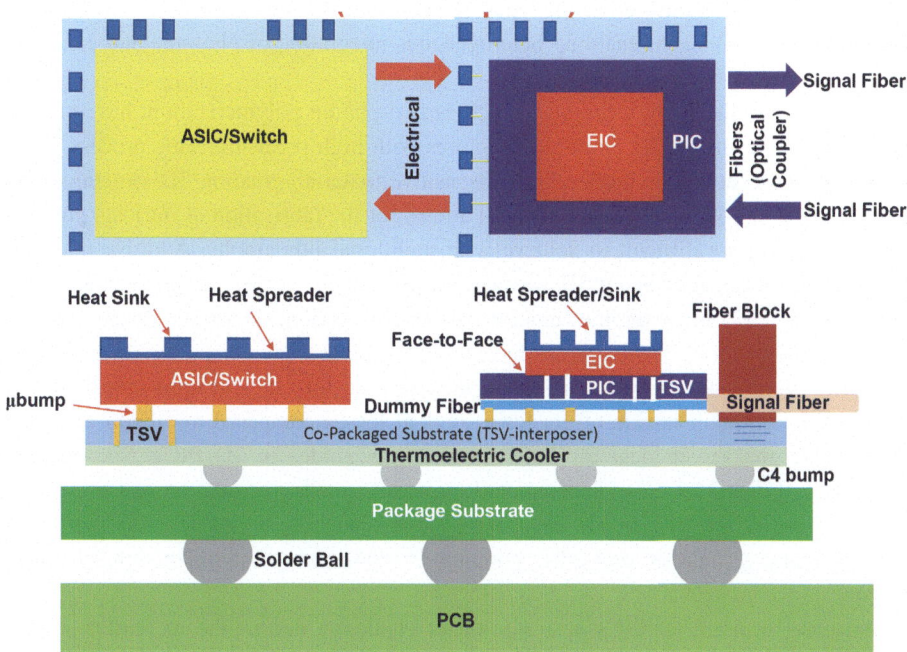

Fig. 3.2 Diagram showing the stacking of layers in 3D PICs, highlighting increased density and performance. 3D heterogeneous integration of ASIC, EIC and PIC on a co-packaged substrate (TSV interposer)

Advanced Fabrication Techniques

Through-Silicon Vias (TSVs) are a critical innovation enabling 3D integration. TSVs are vertical interconnects that pass through silicon wafers, allowing for electrical and optical connections between stacked layers. TSV fabrication involves etching deep holes into the silicon wafer, followed by filling these holes with conductive material, typically copper. Advances in etching techniques and materials deposition have made it possible to create high-aspect-ratio TSVs with low resistance and high reliability. These are used to create vertical interconnections in 3D integrated circuits, enabling the stacking of multiple photonic and electronic layers. This vertical stacking reduces interconnection lengths and improves signal integrity.

Wafer bonding techniques have advanced significantly, enabling the precise alignment and attachment of multiple wafers to form 3D structures. Direct bonding involves joining two wafers without any intermediate layer, relying on molecular adhesion at the interface. Advances in surface preparation and cleanliness have enhanced the strength and reliability of direct bonds. Adhesive bonding uses a thin layer of adhesive material to bond wafers. Innovations in adhesive materials have improved the thermal and mechanical stability of these bonds, making them suitable for high-performance applications. Wafer bonding is essential for integrating different types of materials, such as silicon and III–V semiconductors, in a 3D stack, enabling heterogeneous integration of photonic and electronic components.

Additive manufacturing techniques, such as two-photon polymerization, have enabled the creation of complex 3D photonic structures with high precision. This process uses a focused laser beam to polymerize a photosensitive material, creating 3D structures with resolutions down to the nanometer scale. It allows for the fabrication of intricate photonic components that are difficult to achieve with traditional lithography. Additive manufacturing is used to create custom 3D waveguides, photonic crystals, and other components directly on a chip, enhancing the functionality and integration density of PICs.

The development of new materials has played a crucial role in enabling 3D integration in PICs. Materials with low optical losses are essential for efficient light propagation in 3D structures. Silicon nitride offers lower optical losses compared to silicon at certain wavelengths, making it ideal for long-distance waveguides in 3D PICs. Silicon dioxide, commonly used as a cladding material, provides excellent insulation and low losses, enhancing signal integrity in 3D structures. These materials are used in the fabrication of waveguides, resonators, and other photonic components to minimize signal loss and improve performance.

Managing heat in 3D PICs is a significant challenge due to the increased packing density. Materials with high thermal conductivity help dissipate heat effectively. Diamond has exceptional thermal conductivity and is being explored for use in heat spreaders and substrates in 3D PICs. Graphene's high thermal conductivity and electrical properties make it a promising material for thermal management and electronic interconnects in 3D integrated circuits. These materials are integrated into 3D PICs to enhance thermal management, enabling higher performance and reliability.

III–V semiconductors, such as indium phosphide (InP) and gallium arsenide (GaAs), offer superior optoelectronic properties compared to silicon. Advances in bonding and epitaxial growth techniques have enabled the integration of III–V materials with silicon substrates, combining the best properties of both. These semiconductors are used to fabricate high-efficiency lasers, modulators, and detectors in 3D PICs, enhancing their overall functionality.

Innovations in design methodologies have facilitated the transition to 3D PICs, enabling more efficient and scalable designs. Advanced Electronic Design Automation (EDA) tools have been developed to handle the complexities of 3D integration, providing comprehensive design, simulation, and verification capabilities. 3D layout tools allow designers to create and visualize 3D layouts of PICs, considering the vertical stacking of components and interconnections. Advanced simulation tools model the optical and thermal performance of 3D PICs, enabling optimization of design parameters for maximum efficiency and performance. These tools are used in the design of complex 3D PICs for telecommunications, data centers, and other high-tech applications, ensuring that designs meet stringent performance and reliability requirements.

Process Design Kits (PDKs) provide standardized design rules and models for fabricating PICs, ensuring consistency and compatibility across different fabrication processes. 3D PDKs, developed specifically for 3D integration, include models and design rules for TSVs, wafer bonding, and other 3D-specific processes. These kits streamline the design and fabrication of 3D PICs, reducing development time and costs while improving yield and performance.

The packaging of 3D PICs poses unique challenges, requiring innovative approaches to ensure performance and reliability. Multi-Chip Modules (MCMs) integrate multiple chips, including photonic and electronic components, into a single package, enhancing overall functionality and performance. Advances in hybrid integration techniques have enabled the combination of different types of chips within a single module, leveraging the strengths of each. MCMs are used in high-performance computing, telecommunications, and sensing applications, where they provide compact, high-density solutions.

Efficient optical interconnects are crucial for linking different layers and components in 3D PICs. Innovations in vertical coupling techniques, such as grating couplers and vertical cavity surface-emitting lasers (VCSELs), have improved the efficiency and alignment tolerance of optical interconnects. These interconnects are used in data centers and telecommunications to achieve high-speed, low-latency communication between different layers and components in 3D PICs.

Several emerging technologies are poised to further advance 3D integration in PICs, pushing the boundaries of what is possible. Quantum photonics leverages the principles of quantum mechanics to create new functionalities in PICs. Advances in single-photon sources and detectors have enabled the development of quantum photonic circuits with applications in secure communication and quantum computing. Developing quantum-compatible interconnects and integrating them into 3D PICs is an ongoing area of research, promising to revolutionize information processing and communication. Quantum photonics in 3D PICs has the potential to significantly enhance the performance and capabilities of quantum computers and secure communication networks.

3D PICs are increasingly being used for integrated sensing applications, combining multiple sensing modalities on a single chip. Combining optical, electrical, and mechanical sensing elements in a 3D structure enhances the sensitivity and functionality of sensors. Integrated sensing in 3D PICs is used in medical diagnostics, environmental monitoring, and industrial inspection, providing compact, high-performance solutions.

The transition from 2D to 3D integration in Photonic Integrated Circuits represents a significant technological advancement, enabled by breakthroughs in fabrication techniques, materials, design methodologies, and packaging technologies. These innovations have addressed many of the limitations of 2D PICs, such as integration density, interconnection length, and performance, paving the way for more complex and capable photonic systems. As research and development continue, 3D PICs are expected to drive further advancements in telecommunications, data centers, quantum computing, and beyond, unlocking new possibilities and applications.

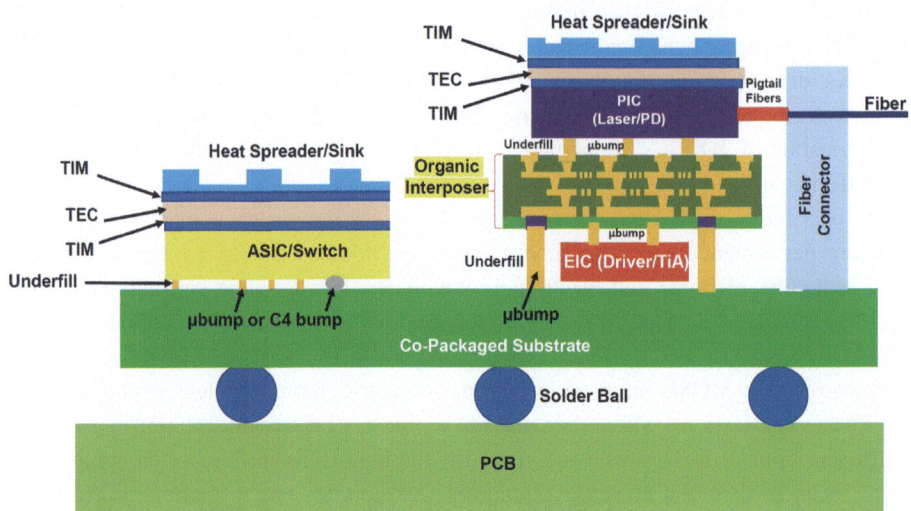

Fig. 3.3 Case study example illustrating performance improvements achieved through 3D PICs. Another Co-packaged optics method for 51.2 Tbit/s switch

3.3 Case Studies: Examples of Enhanced Performance with 3D PICs Compared to Their 2D Counterparts

The transition from 2D to 3D Photonic Integrated Circuits (PICs) has brought about significant improvements in performance across various applications. This chapter delves into several case studies that illustrate the enhanced performance of 3D PICs compared to their 2D counterparts, highlighting advancements in telecommunications, data centers, sensing, and quantum computing (Fig. 3.3).

Case Study 1: Telecommunications

Telecommunications networks rely heavily on PICs to transmit and process data at high speeds. Traditional 2D PICs have served this purpose well, but they face limitations in integration density, interconnection length, and thermal management, which can hinder performance.

2D PICs have limited integration density, constraining the number of components that can be packed on a single chip. This limitation affects the overall capacity and functionality of the network components. Longer interconnections in 2D PICs lead to higher propagation losses and increased latency, impacting signal integrity and speed. Additionally, the dense packing of components in 2D planes leads to challenges in dissipating heat, which can degrade performance and reliability.

By stacking components vertically, 3D PICs significantly increase integration density. This vertical stacking reduces the footprint of the devices while enhancing their functionality and capacity. 3D integration reduces the distance between components, minimizing propagation losses and latency. This improvement leads to higher data transmission speeds and better signal integrity. Furthermore, 3D PICs distribute heat more effectively across multiple layers, reducing thermal hotspots and improving overall thermal performance.

Performance Enhancement
A specific example of enhanced performance in telecommunications is the advancement of 3D Photonic Integrated Circuit (PIC)-based transceivers for fiber-optic networks. These cutting-edge transceivers employ vertical integration, meticulously combining lasers, modulators, and detectors into a single, compact, high-density package. This integration not only optimizes space utilization but also leads to substantial improvements in data transmission rates, enabling networks to handle higher volumes of data with increased efficiency. Compared to their 2D PIC-based counterparts, 3D PIC transceivers exhibit markedly reduced latency, ensuring faster and more reliable communication, which is crucial for applications demanding real-time data processing and transmission. Furthermore, the sophisticated design of 3D PIC transceivers incorporates advanced thermal management techniques. By effectively dissipating heat, these transceivers maintain stable operation even at elevated power levels, which is essential for sustaining high-performance metrics over extended periods. Moreover, the implementation of 3D PIC technology in fiber-optic networks contributes to overall system robustness and reliability. The enhanced thermal management not only prolongs the lifespan of the transceivers but also minimizes the risk of performance degradation over time. This technological leap forward supports the growing demands of modern telecommunications infrastructure, accommodating the exponential increase in data traffic driven by innovations such as 5G, the Internet of Things (IoT), and high-definition streaming services. In addition to performance benefits, 3D PIC-based transceivers offer economic advantages by reducing the need for multiple discrete components and simplifying the manufacturing process. The integrated nature of these devices lowers production costs and enhances scalability, making it feasible to deploy high-performance fiber-optic networks on a broader scale. This convergence of performance and cost-efficiency positions 3D PIC technology as a cornerstone in the future development of telecommunications, paving the way for faster, more reliable, and more efficient global communication networks.

Case Study 2: Data Centers
Data centers are the backbone of the digital economy, requiring high-speed, efficient data processing and transmission. PICs play a critical role in enabling optical interconnects within data centers, where performance and scalability are paramount.

The limited integration density of 2D PICs restricts the number of optical interconnects that can be implemented, affecting the data throughput and scalability of data center networks. Longer interconnections in 2D PICs lead to higher signal loss and increased latency, reducing the efficiency of data transfer. Additionally, the power consumption associated with driving long interconnections in 2D PICs contributes to higher operational costs and thermal management challenges.

3D PICs allow for the integration of more optical interconnects within the same footprint, increasing the overall data throughput and scalability of data center networks. The vertical stacking of components in 3D PICs shortens interconnection lengths, reducing signal loss and latency, and enhancing data transfer efficiency. By minimizing interconnection lengths, 3D PICs reduce the power required for signal transmission, leading to lower operational costs and improved energy efficiency.

Performance Enhancement

An example of the enhanced performance enabled by 3D Photonic Integrated Circuit (PIC)-based optical switches can be observed in modern data center networks. As data traffic continues to grow exponentially due to cloud services, streaming, and AI-driven applications, data centers are under immense pressure to provide faster, more scalable, and energy-efficient solutions for managing this data flow. 3D PICs offer an innovative solution by leveraging their unique vertical integration to overcome the limitations of traditional 2D PIC-based switches.

Ultra-fast Switching Speeds and Low Latency

The high integration density provided by 3D PICs allows for the stacking of multiple optical components, such as waveguides, modulators, and detectors, in a compact vertical arrangement. This vertical stacking not only reduces the physical footprint but also minimizes the length of interconnects between components. With shorter optical paths between the switching elements, 3D PIC-based switches achieve ultra-fast switching speeds, reducing data packet delays as signals move through the network. This is crucial for real-time data processing and low-latency applications, such as AI training and big data analytics. By optimizing the signal flow, 3D optical switches drastically lower latency compared to their 2D counterparts, making them ideal for high-frequency trading, gaming, and cloud computing services, where milliseconds can impact performance.

Increased Data Throughput

Due to the vertical stacking of layers in 3D PICs, these switches are capable of handling a significantly higher volume of data. The compact nature of 3D integration allows for more switching elements to be placed on the same chip, which translates to a greater number of parallel optical channels. The ability to integrate multiple switching components vertically in a small area enables 3D PIC switches to support a higher number of data streams simultaneously, effectively increasing data throughput without expanding the physical footprint of the switch. Many 3D PIC-based switches also take advantage of Wavelength Division Multiplexing (WDM) technology, where multiple data streams are transmitted on different wavelengths over a single optical fiber. The enhanced integration of components in 3D PICs allows for more precise control of these wavelengths, further boosting throughput.

Lower Power Consumption

Power consumption is a critical factor in data center operations, as cooling and energy costs constitute a large portion of the overall expenses. Traditional 2D PICs face challenges related to heat dissipation due to the relatively large distances between interconnects and components. 3D PIC-based optical switches offer a more energy-efficient solution because the shorter interconnections and reduced signal attenuation lead to lower power requirements for signal transmission. With the components closer together in a vertical stack, less energy is needed to drive the signals through the switching elements, cutting overall power consumption. By stacking components vertically, 3D PICs also benefit from more effective heat distribution strategies. Advanced cooling systems and thermal management materials can be incorporated into the layers, allowing for better dissipation of heat, which further reduces power consumption and the need for extensive cooling infrastructures.

Scalability for Future Data Center Demands

Data centers must continually scale to meet growing demands, and 3D PIC-based optical switches offer a solution that can keep pace with future requirements. 3D integration allows for higher packing density without significantly increasing the physical size of the switching units, making it easier for data centers to scale operations without requiring additional space. This is especially advantageous in urban data centers, where real estate is at a premium. 3D PIC-based switches can be designed to seamlessly integrate with existing optical infrastructure, offering an upgrade path for data centers looking to enhance performance without completely overhauling their network architecture.

Comparison with 2D PIC-Based Switches

Compared to 2D PIC-based switches, the 3D variants offer significant improvements in data throughput and latency. 2D PICs are limited by the horizontal plane, leading to longer interconnection paths, higher power consumption, and scalability challenges. 3D PIC switches consume less power due to their optimized design, reducing the energy required to drive signals and decreasing overall operational costs for data centers. Additionally, 3D PICs can support a higher number of parallel data streams and use WDM technology more effectively, making them a superior choice for future-proofing data centers.

The implementation of 3D PIC-based optical switches marks a significant leap forward in data center technology. By leveraging the high integration density, shorter interconnections, and energy efficiency of 3D PICs, these switches enable ultra-fast switching speeds, low latency, and high data throughput while reducing power consumption. These advancements not only improve current data center operations but also provide a scalable, cost-effective solution to meet the growing demands of cloud computing, telecommunications, and high-performance computing.

Case Study 3: Sensing
PICs are widely used in sensing applications, including environmental monitoring, medical diagnostics, and industrial inspection. The ability to integrate multiple sensing modalities on a single chip is crucial for enhancing sensitivity and functionality.

2D PICs have limited space for integrating multiple sensing elements, restricting the complexity and functionality of the sensors. Closely packed components in 2D PICs can lead to crosstalk and signal interference, reducing the accuracy and sensitivity of the sensors. Additionally, the heat generated by densely packed components in 2D PICs can affect the stability and reliability of the sensors.

3D PICs enable the integration of multiple layers of sensing elements, increasing the complexity and functionality of the sensors without enlarging the footprint. The separation of components across different layers in 3D PICs reduces crosstalk and signal interference, enhancing the accuracy and sensitivity of the sensors. Distributing heat across multiple layers in 3D PICs enhances thermal management, ensuring stable and reliable sensor operation.

A notable example of enhanced performance in sensing is the development of 3D PIC-based biosensors for medical diagnostics. These biosensors integrate multiple layers of photonic elements, such as waveguides, resonators, and detectors, to achieve high sensitivity and specificity in detecting biomarkers. Compared to 2D PIC-based biosensors, the 3D versions offer improved performance due to reduced crosstalk, enhanced integration density, and better thermal management, leading to more accurate and reliable diagnostic results.

Case Study 4: Quantum Computing

Background
Quantum computing leverages the principles of quantum mechanics to perform complex computations that are beyond the capabilities of classical computers. PICs are essential for implementing quantum photonic circuits, which require precise control of single photons and quantum states.

2D PICs have limited integration density, restricting the number of quantum photonic elements that can be integrated, which affects the complexity and scalability of quantum circuits. Longer interconnections in 2D PICs lead to higher photon loss, reducing the efficiency of quantum operations. Additionally, closely packed quantum elements in 2D PICs can cause crosstalk and interference, degrading the fidelity of quantum states.

3D PICs enable the integration of more quantum photonic elements within the same footprint, increasing the complexity and scalability of quantum circuits. Vertical integration in 3D PICs reduces interconnection lengths, minimizing photon loss and enhancing the efficiency of quantum operations. The separation of quantum elements across different layers in 3D PICs reduces crosstalk and interference, improving the fidelity of quantum states.

Performance Enhancement

An example of enhanced performance in quantum computing is the development of 3D PIC-based quantum photonic processors. These processors integrate multiple layers of waveguides, beam splitters, and detectors to perform complex quantum operations. Compared to 2D PIC-based quantum processors, the 3D versions offer higher qubit counts, improved fidelity, and greater operational efficiency, paving the way for more powerful and scalable quantum computers (Fig. 3.4).

Conclusion

The transition from 2D to 3D Photonic Integrated Circuits (PICs) marks a transformative leap in the performance and functionality of photonic systems, influencing a wide range of advanced applications. The limitations of 2D PICs—such as restricted integration density, longer interconnect paths, and challenges in managing heat and signal integrity—are effectively addressed by the innovations inherent to 3D PIC technology. Here's a more detailed exploration of the advancements and their impact on various fields:

Higher Integration Density

One of the most significant advantages of 3D PICs is their ability to integrate multiple photonic and electronic components vertically, utilizing the third dimension for enhanced scalability. This higher integration density enables more compact designs with a larger number of functional components stacked within the same chip footprint. For telecommunication systems, 3D PICs allow for a higher number of modulators, filters, and multiplexers to be integrated into a single chip, improving data transmission rates and channel capacity. This increased density is particularly beneficial in wavelength-division multiplexing (WDM) systems, where multiple channels of data are carried over a single optical fiber. As the demand for bandwidth and data processing grows, especially with the expansion of 5G networks, 3D PICs offer a pathway for future scalability without significant increases in size or cost.

Shorter Interconnections

By stacking components vertically, 3D PICs reduce the length of interconnects between them, leading to faster signal propagation and lower power consumption. Shorter optical paths reduce delays and latency, enabling real-time data processing and ultra-fast data transmission. This is crucial in applications such as data centers, where millions of packets are transferred across networks every second. The reduced interconnection length decreases signal loss and energy dissipation, cutting down on overall power consumption.

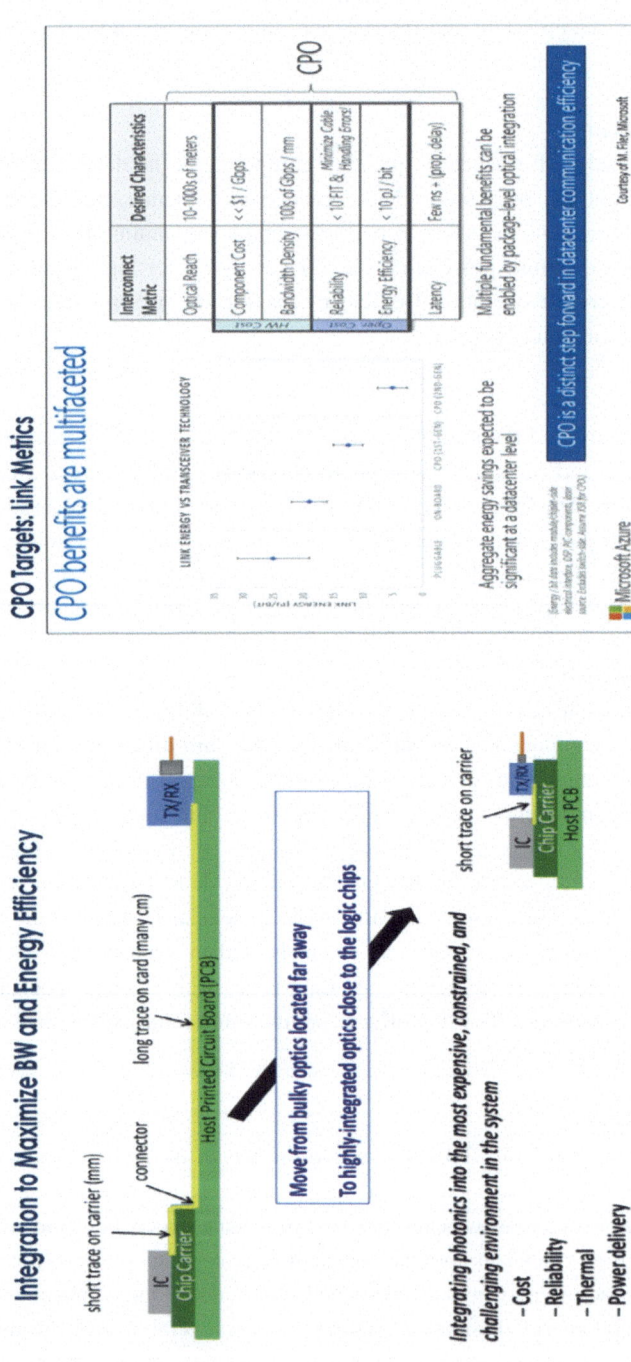

Fig. 3.4 Co-packaged photonics for improved energy efficiency and performance of AI applications. C. Schow IEEE International Electron Devices Meeting (IEDM) Tutorial, Dec. 2024

3.3 Case Studies: Examples of Enhanced Performance with 3D PICs ...

This is especially important for high-performance computing (HPC) environments, where efficiency and speed must be balanced with sustainability.

Improved Thermal Management
In traditional 2D PICs, densely packed components generate heat, which can degrade performance and reliability over time. 3D PICs address this issue by spreading components across multiple layers, allowing for more efficient heat dissipation and thermal management. The 3D architecture provides more surface area and opportunities for incorporating thermal management materials and cooling systems. Microfluidic cooling systems, for example, can be integrated into 3D PICs to actively manage heat, making them suitable for applications where reliability and longevity are critical, such as optical communication networks and supercomputing. With better heat management, 3D PICs can operate at higher power levels without compromising performance, which is essential for solid-state LiDAR systems used in autonomous vehicles and robotics.

Reduced Signal Interference
In 2D PICs, components placed too closely together can cause unwanted crosstalk and signal interference, which limits their performance, especially in high-frequency applications. 3D PICs mitigate this issue by allowing components to be isolated across different layers, reducing signal interference and improving overall signal integrity. In sensing applications, such as biosensing or quantum computing, precise control over signal integrity is critical. By reducing crosstalk, 3D PICs provide cleaner, more accurate data in optical sensing and spectroscopy applications. Quantum PICs stand to benefit immensely from 3D integration, as quantum information processing requires extremely low interference and high coherence to accurately perform calculations.

Applications in Quantum Computing and Sensing
The advancements brought by 3D PICs are particularly promising for fields such as quantum computing and advanced sensing. 3D integration allows for more complex quantum photonic circuits, enabling the realization of quantum gates and quantum entanglement on a scalable platform. The ability to integrate multiple quantum components on a single chip will drive the development of quantum processors and quantum communication networks. In applications such as LiDAR for autonomous vehicles or biosensors for medical diagnostics, the compactness, efficiency, and reduced power consumption of 3D PICs allow for more accurate and faster sensing capabilities, even in resource-constrained environments.

Ongoing Research and Future Prospects
As research and development continue, 3D PICs are expected to push the boundaries of photonics even further. Current research efforts focus on materials innovation, such as integrating new materials like silicon carbide or graphene into 3D PICs, which could unlock new capabilities such as enhanced thermal conductivity, higher power handling,

or even the development of plasmonic circuits that manipulate light at the nanoscale. The development of standardized process design kits (PDKs) and automated design tools for 3D PICs will streamline the design and fabrication processes, making them more accessible to a wider range of industries. As fabrication techniques for 3D PICs mature, their cost is expected to decrease, enabling broader adoption in commercial applications such as consumer electronics, healthcare, and IoT (Internet of Things) devices.

Revolutionizing the Future

The transition from 2D to 3D PICs is not just an incremental improvement—it is a fundamental shift that will revolutionize telecommunications, data processing, sensing technologies, and quantum computing. By addressing the limitations of traditional 2D architectures, 3D PICs enable more efficient, scalable, and powerful solutions that are critical to the next wave of technological innovation. As the field continues to evolve, the impact of 3D PICs will likely expand, unlocking new possibilities in photonics and driving innovation in industries that rely on cutting-edge optical technologies.

Design and Fabrication Techniques

4

- **CMOS-Compatible Processes**: Integration of photonics with mature CMOS fabrication techniques.
- **Additive Manufacturing in Photonics**: Techniques like two-photon polymerization for creating intricate 3D structures.
- **Advanced Lithography and Etching Methods**: Detailed processes for precise fabrication.

4.1 CMOS-Compatible Processes: Integration of Photonics with Mature CMOS Fabrication Techniques

The integration of photonic devices with CMOS (Complementary Metal-Oxide-Semiconductor) fabrication techniques has revolutionized the field of Photonic Integrated Circuits (PICs) by enabling the mass production of photonic components using the same processes that have been refined for decades in the electronics industry. This compatibility with established CMOS processes has opened up new opportunities for the large-scale deployment of PICs, driving advancements in areas like optical communications, data centers, and sensing technologies. By leveraging CMOS fabrication, PICs can now be produced with the precision, scalability, and cost-effectiveness required for commercial applications. However, integrating photonics with CMOS processes presents unique challenges, such as managing material compatibility, ensuring low optical loss, and optimizing device performance in a traditionally electronic-centric environment. Breakthroughs in silicon photonics—particularly in the development of efficient waveguides, modulators, and photodetectors—have addressed many of these challenges, allowing for the seamless coexistence of photonic and electronic components on the same chip. Additionally, advances in 3D integration, wafer bonding, and heterogeneous integration have expanded the potential of CMOS-compatible PICs, enabling more complex and multifunctional systems that combine the best of both photonics and electronics. This chapter will delve into these technological breakthroughs and explore how they are shaping the future of integrated photonic systems.

Overview of CMOS Technology
See Fig. 4.1.

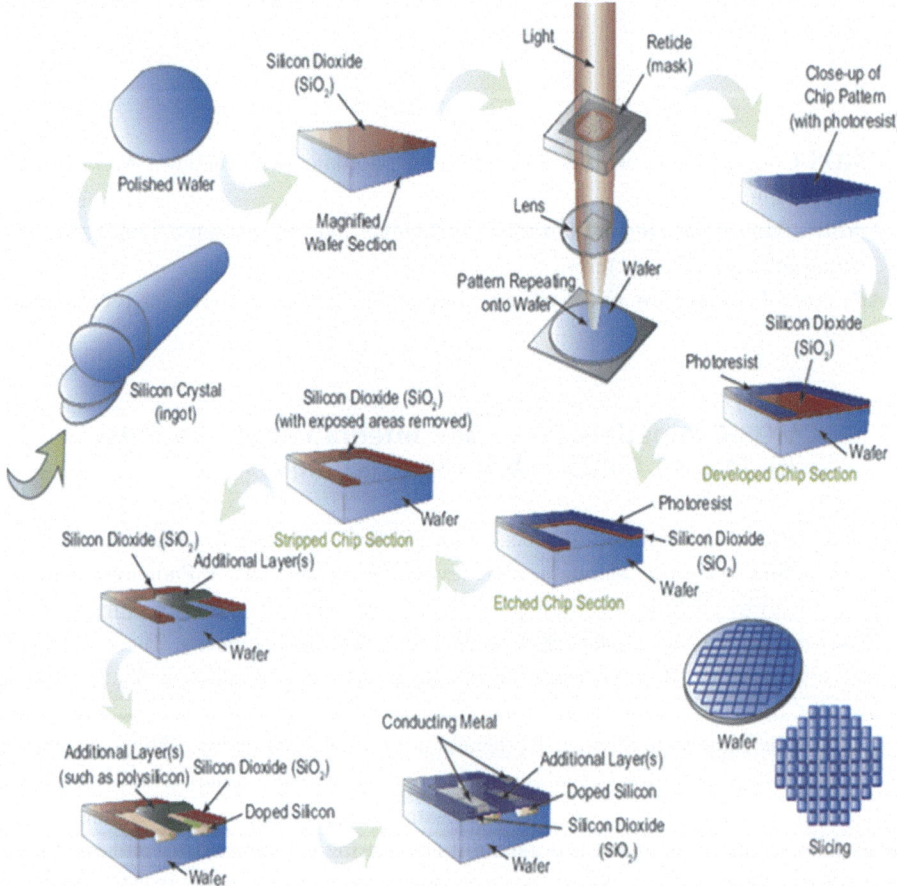

Fig. 4.1 Flowchart of CMOS-compatible fabrication processes, showing each step in the integration of photonics

4.1 CMOS-Compatible Processes: Integration of Photonics ...

CMOS Fabrication Basics

CMOS technology is the foundation of modern electronics, widely used for manufacturing integrated circuits (ICs) due to its low power consumption, high noise immunity, and scalability. The fabrication process begins with substrate preparation, where a high-purity silicon wafer is cleaned and doped to create regions with different electrical properties. Layer deposition follows, in which thin layers of materials such as silicon dioxide (SiO_2) and polysilicon are applied using techniques like chemical vapor deposition (CVD) and physical vapor deposition (PVD). Lithography is then employed to pattern these layers, defining the geometry of transistors and interconnects. Etching removes unwanted material to shape the desired structures, while doping through ion implantation or diffusion modifies the electrical properties of specific regions. Finally, metallization adds metal layers to form electrical interconnections between transistors.

CMOS Advantages

CMOS technology offers several advantages, including scalability, allowing millions of transistors to be integrated on a single chip, enabling complex and powerful circuits. Its cost-effectiveness benefits from economies of scale, reducing manufacturing costs, while its reliability ensures consistent performance through well-established fabrication techniques.

Integration of Photonics with CMOS

Integrating photonic devices with CMOS technology involves incorporating photonic components such as waveguides, modulators, and detectors into the existing CMOS process. This hybrid integration enables the development of photonic-electronic circuits that leverage the strengths of both technologies.

CMOS-Compatible Photonic Components

Waveguides are essential for guiding light on a chip using total internal reflection. Silicon waveguides are commonly used due to their compatibility with CMOS technology, while silicon nitride (SiN) waveguides offer lower optical losses for specific wavelengths. Modulators integrate electro-optic effects to control light signals. Silicon-based modulators leverage changes in refractive index induced by an applied electric field, enabling signal modulation. Detectors, particularly germanium photodetectors, efficiently convert near-infrared light into electrical signals, ensuring seamless integration with silicon photonics.

Process Integration Steps

The integration process often begins with Silicon-On-Insulator (SOI) wafers, which consist of a thin silicon layer on an insulating SiO_2 layer atop a silicon substrate. These wafers serve as the foundation for photonic components. Photonic layer patterning is achieved through photolithography and etching, defining structures such as waveguides and modulators with high precision. Doping and implantation introduce dopants into specific regions to create p-n junctions in modulators and photodetectors, mirroring steps in traditional CMOS fabrication. Metallization follows, forming electrical contacts for photonic components and ensuring proper electrical interfacing with CMOS circuits. Planarization using chemical-mechanical polishing (CMP) ensures a smooth wafer surface, which is essential for maintaining component performance and reliability.

Challenges and Solutions
Thermal budget considerations arise because photonic components can be sensitive to high temperatures. Careful sequencing of fabrication steps and low-temperature processing techniques mitigate performance degradation. Material compatibility is another challenge, as integrating materials like germanium for photodetectors and silicon nitride for waveguides requires advanced deposition and interface engineering techniques to prevent defects. Alignment precision is critical for efficient coupling and signal integrity, necessitating advanced lithography, real-time monitoring, and feedback systems.

Breakthroughs in CMOS-Compatible Photonics

Monolithic Integration
Monolithic integration involves fabricating photonic and electronic components on the same silicon chip, eliminating external assembly steps. This approach enhances compactness, reducing device footprints while increasing integration density. It improves performance by minimizing insertion losses and enhances manufacturing efficiency by streamlining processes. A key application of monolithic integration is silicon photonic transceivers for data communication, where lasers, modulators, waveguides, and detectors are integrated onto a single chip, achieving high-speed transmission with low power consumption.

Heterogeneous Integration
Heterogeneous integration combines different material systems, such as III–V semiconductors and silicon, to create hybrid photonic-electronic circuits. This approach leverages the superior optoelectronic properties of III–V materials while maintaining CMOS scalability. Direct bonding techniques align and bond III–V materials to silicon substrates at the atomic level, creating seamless interfaces. Selective area growth enables precise placement of III–V materials on predefined silicon regions, optimizing their performance. Heterogeneous integration has enabled the development of hybrid silicon lasers and high-speed modulators, which combine the strengths of both material systems.

Advanced Packaging
Efficient packaging techniques ensure seamless integration of photonic and electronic components while maintaining performance and reliability. Flip-chip bonding attaches photonic chips face-down onto electronic substrates using solder bumps or conductive

adhesives, minimizing optical interconnect lengths and improving mechanical stability. Wafer-level packaging encapsulates photonic and electronic components at the wafer scale, enhancing protection and facilitating mass production, reducing manufacturing costs and improving reliability.

Future Prospects

The continued advancement of CMOS-compatible photonics is driven by innovations in materials, design methodologies, and packaging techniques. The integration of new materials, such as silicon carbide or graphene, promises improved thermal conductivity, enhanced power handling, and the potential for nanoscale plasmonic circuits. Standardized process design kits (PDKs) and automated design tools will streamline fabrication, making photonic integration more accessible across industries. As fabrication techniques mature, cost reductions will facilitate widespread adoption, unlocking new applications in high-speed communications, medical diagnostics, and next-generation computing.

Applications of CMOS-Compatible Photonics

Data Centers

The integration of photonic components with CMOS technology has revolutionized data centers, enabling high-speed, low-power optical interconnects that enhance data transfer efficiency and reduce latency. CMOS-compatible photonic transceivers and switches are now widely used in data center networks, supporting the ever-increasing demand for data throughput and bandwidth.

Telecommunications

CMOS-compatible photonics has significantly impacted telecommunications, providing high-performance components for fiber-optic networks. Silicon photonic transceivers, modulators, and detectors integrated with CMOS technology enable high-speed data transmission over long distances, supporting the growth of high-speed internet and 5G networks.

Medical Diagnostics

In medical diagnostics, CMOS-compatible photonic sensors offer high sensitivity and specificity for detecting biomarkers and monitoring physiological parameters. These sensors integrate photonic components with CMOS electronics to provide compact, cost-effective solutions for point-of-care diagnostics and wearable health monitoring devices.

Sensing and Imaging

CMOS-compatible photonics is also used in various sensing and imaging applications, including environmental monitoring, industrial inspection, and autonomous vehicles. Integrated photonic sensors provide high-resolution, real-time data, enabling advanced sensing capabilities in compact and robust packages.

Conclusion

The integration of photonic devices with mature CMOS fabrication techniques has paved the way for advanced Photonic Integrated Circuits that combine the best of photonics and electronics. CMOS-compatible processes have enabled the development of compact, high-performance, and cost-effective photonic devices, driving innovation in telecommunications, data centers, medical diagnostics, and beyond. As research and development continue to advance, CMOS-compatible photonics is expected to further revolutionize various fields, unlocking new possibilities and applications.

4.2 Additive Manufacturing in Photonics: Techniques Like Two-Photon Polymerization for Creating Intricate 3D Structures

Additive manufacturing, commonly known as 3D printing, has dramatically transformed the fabrication of intricate structures across various industries, and its impact on photonics is particularly profound. In the field of photonics, where precision and complexity are paramount, additive manufacturing techniques—especially two-photon polymerization (2PP)—have unlocked new possibilities for creating three-dimensional micro- and nanostructures with unparalleled accuracy. Two-photon polymerization works by focusing a laser beam to initiate polymerization only at the precise focal point, allowing for the fabrication of highly detailed structures at the nanoscale that would be impossible using traditional manufacturing techniques. This level of control has made 2PP a powerful tool in the development of photonic components such as optical waveguides, metasurfaces, and photonic crystals. Additionally, 2PP is being used to create optical interconnects and microlenses, enabling new approaches to optical data transmission, imaging, and sensing. The ability to build complex 3D structures layer by layer has revolutionized the design of photonic devices, allowing for highly customizable geometries and novel functionalities that can enhance device performance. This chapter will delve into the principles of additive manufacturing in photonics, highlighting the applications of two-photon polymerization and the transformative impact these technologies are having on the design, fabrication, and integration of advanced photonic systems (Fig. 4.2).

Fig. 4.2 Illustration of additive manufacturing in photonics, highlighting two-photon polymerization to create 3D structures. https://www.microlight3d.com/technology/two-photon-polymerization

Overview of Additive Manufacturing

Additive manufacturing builds objects layer by layer from a digital model, allowing for the creation of complex geometries that are difficult or impossible to achieve with traditional subtractive manufacturing methods. In photonics, additive manufacturing provides significant advantages, including the ability to create intricate designs that enhance device functionality, customized fabrication tailored to specific applications, and rapid prototyping that accelerates development cycles and design iterations.

Two-Photon Polymerization (2PP)

Two-photon polymerization (2PP) is a high-resolution additive manufacturing technique that uses a focused laser beam to polymerize a photosensitive material at the focal point, enabling the fabrication of 3D structures with nanoscale precision.

Principles of Two-Photon Polymerization

2PP relies on two-photon absorption, where two photons are absorbed simultaneously by a photosensitive material to initiate polymerization. This process occurs only at the laser's focal point, ensuring precise control over polymerization. Since it is a nonlinear optical process, polymerization is confined strictly to the focal region, preventing unwanted exposure and allowing for highly detailed 3D structures. The fabrication proceeds layer by layer, where the laser scans a predetermined pattern to build up the structure in a controlled and precise manner.

Process Parameters

Several key parameters influence the effectiveness of 2PP. The laser wavelength must be carefully chosen to optimize two-photon absorption, with near-infrared lasers commonly used to balance efficiency and minimize material damage. The photosensitive material must exhibit a high two-photon absorption cross-section while maintaining suitable mechanical and optical properties post-polymerization. Scanning speed and power also play a critical role; higher laser power increases polymerization rates but can reduce precision, while slower scanning improves resolution at the cost of longer fabrication times.

Applications of Two-Photon Polymerization in Photonics
2PP has enabled the development of highly functional photonic devices and microstructures across various applications.

Micro-optics

2PP allows for the fabrication of micro-lenses with precise curvature and smooth surfaces, which are crucial for focusing and directing light in photonic circuits. It also enables the production of diffractive optical elements, such as gratings and holograms, for advanced light manipulation and beam shaping.

Photonic Crystals

Using 2PP, 3D photonic crystals can be fabricated with highly controlled periodic structures that create photonic bandgaps, improving waveguides, optical filters, and resonators. Custom geometries can be designed to achieve specific optical properties, optimizing photonic devices for specialized applications.

Waveguides

2PP is used to create embedded waveguides, which allow for precise control of light paths with minimal losses, a critical requirement for integrated photonic circuits. Both multimode and single-mode waveguides can be fabricated, offering flexibility in circuit design and performance optimization.

Sensing and Detection

Biosensors produced using 2PP feature intricate microstructures that enhance sensitivity and specificity, making them valuable in detecting biological molecules, pathogens, and environmental contaminants. Optical sensors with customized geometries can be tailored for applications in environmental monitoring, industrial inspection, and medical diagnostics.

Advantages of Two-Photon Polymerization

2PP offers significant benefits over traditional fabrication techniques, making it a powerful tool for photonic device manufacturing.

High Resolution and Precision

2PP achieves nanoscale feature sizes, enabling the creation of detailed and intricate structures. Its precision control ensures minimal defects and high structural integrity, making it ideal for applications that require exceptional accuracy.

Material Versatility

The technique supports a wide range of materials, including polymers, composites, and hybrid materials, offering flexibility in photonic device design. Functional materials with specific optical, mechanical, or chemical properties can be used to meet diverse application requirements.

Customization and Flexibility

2PP provides unparalleled design freedom, allowing engineers to create complex and custom geometries tailored to specific applications. Its rapid prototyping capabilities enable quick iterations, reducing development time and enhancing design optimization.

Integration with Other Techniques

2PP can be combined with traditional fabrication methods, enabling hybrid manufacturing that leverages the strengths of multiple processes. Advanced multi-material printing techniques allow the simultaneous fabrication of different materials, enabling multifunctional photonic and electronic devices.

Challenges and Future Directions
While 2PP has revolutionized photonic fabrication, several challenges remain.

Scalability

The fabrication speed of 2PP is limited by its layer-by-layer approach, making large-scale production time-intensive. Advances in laser scanning technology and parallelization methods will be needed to improve scalability. Additionally, the cost of 2PP systems and materials remains high, requiring research into more cost effective materials and process optimizations.

Material Properties

Expanding the range of photosensitive materials suitable for 2PP is crucial for broadening its applications. Research into new materials with improved optical and mechanical properties will enhance the versatility of the technique. Post-processing steps, such as curing and cleaning, are often required to achieve optimal material characteristics, and simplifying these steps would improve fabrication efficiency.

Design Complexity

Advanced simulation and modeling tools are needed to accurately predict the performance of complex 3D photonic structures. Improvements in computational methods and software will assist in optimizing 2PP designs, balancing resolution, mechanical stability, and optical performance.

Future Directions
The future of two-photon polymerization in photonics is promising, with ongoing research focusing on enhancing capabilities and overcoming existing challenges.

- High-Speed 2PP: Advances in laser technology and parallel processing techniques will significantly increase fabrication speed, making large-scale production more feasible.
- Multi-Material and Functional Printing: The development of systems capable of printing multiple materials and integrating functional elements will enable more sophisticated and multifunctional photonic devices.
- Integration with Conventional Manufacturing: Combining 2PP with traditional manufacturing techniques will create hybrid fabrication processes that leverage the advantages of both approaches, enhancing photonic device performance.
- New Material Development: Research into advanced photosensitive materials with superior optical, mechanical, and chemical properties will expand the range of 2PP applications, further pushing the boundaries of photonic technology.

Conclusion
Two-photon polymerization represents a groundbreaking advancement in additive manufacturing for photonics, allowing for the creation of intricate 3D structures with unparalleled precision and complexity. This technique has already demonstrated significant potential in applications such as micro-optics, photonic crystals, waveguides, and sensing technologies. While challenges remain in scalability, cost, and material properties, ongoing research and development are poised to further enhance the capabilities and accessibility of 2PP. As innovation continues, 2PP will drive advancements in photonic technology, unlocking new possibilities in data communication, medical diagnostics, and next-generation computing.

4.3 Advanced Lithography and Etching Methods: Detailed Processes for Precise Fabrication

Advanced lithography and etching methods are crucial for the precise and scalable fabrication of photonic integrated circuits (PICs) and other micro- and nanoscale devices. These techniques enable the creation of intricate patterns with sub-micron and nanoscale

4.3 Advanced Lithography and Etching Methods: Detailed ...

resolution, which is essential for the performance, integration, and functionality of modern photonic devices. Lithography, particularly deep ultraviolet (DUV) and electron beam lithography (EBL), allows for the direct patterning of photonic components such as waveguides, resonators, and modulators with extreme accuracy, while also supporting the scaling down of features to meet the demands of ever-smaller and faster devices. Etching processes, including reactive ion etching (RIE) and inductively coupled plasma (ICP) etching, are equally vital, providing the means to sculpt materials like silicon, indium phosphide, and gallium arsenide into precise structures that are critical for guiding and manipulating light at very small scales. Recent advancements in lithography, such as extreme ultraviolet (EUV) lithography, have pushed the boundaries of pattern resolution, enabling even more compact and complex PICs for use in fields such as telecommunications, quantum computing, and medical imaging. Furthermore, hybrid techniques that combine advanced lithography with multi-layer etching have allowed the fabrication of 3D photonic structures, expanding the capabilities of PICs and enabling new functionalities such as integrated optical interconnects and advanced sensing systems. This chapter explores the fundamental principles of these fabrication methods, detailing the innovations and challenges in developing ever-more precise and efficient techniques to meet the growing demands of the photonics industry.

Lithography Methods

Lithography is a crucial process in photonic integrated circuit (PIC) fabrication, enabling the precise transfer of patterns from a mask onto a substrate. This step defines the shapes and positions of various photonic components, ensuring their functionality and integration.

Photolithography

Photolithography is the most widely used lithography technique in semiconductor manufacturing. The process begins with coating, where a light-sensitive photoresist is applied to the substrate's surface. In the exposure stage, ultraviolet (UV) light is directed through a photomask containing the desired pattern, altering the chemical structure of the exposed photoresist. During development, either the exposed or unexposed regions of the resist are washed away, depending on whether a positive or negative resist is used. Finally, the pattern is transferred to the underlying material through etching, completing the process.

Photolithography offers high throughput and precision, making it ideal for mass production. However, its resolution is limited by the wavelength of UV light, typically around 193 nm, which restricts feature sizes to approximately 50 nm.

Electron Beam Lithography (EBL)

Electron Beam Lithography (EBL) employs a focused beam of electrons to write patterns directly onto a substrate coated with an electron-sensitive resist. Unlike photolithography, EBL does not require a mask, allowing for direct writing and greater flexibility in pattern design. The high resolution of EBL, reaching a few nanometers, is due to the shorter wavelength of electrons. However, EBL suffers from low throughput, as it writes patterns sequentially rather than in parallel, making it more suitable for research and prototyping rather than large-scale production.

Nanoimprint Lithography (NIL)
Nanoimprint Lithography (NIL) is a high-resolution, cost-effective technique that transfers patterns using a mold. The process involves imprinting, where a hard mold with the desired pattern is pressed into a soft resist on the substrate, followed by curing using UV light or heat. Once the mold is removed, the pattern remains imprinted on the resist, and etching transfers it into the substrate. NIL provides high resolution and low-cost fabrication, making it suitable for research and commercial applications, though the challenge lies in producing high-quality molds.

Extreme Ultraviolet Lithography (EUVL)
Extreme Ultraviolet Lithography (EUVL) employs a much shorter wavelength of light (around 13.5 nm) to achieve extremely fine patterning. The higher resolution enables feature sizes down to a few nanometers, making EUVL a key technology for next-generation semiconductor and photonic devices. However, it requires complex and expensive equipment, including specialized EUV light sources and precise optics for focusing and directing the light.

Etching Methods
Etching removes material from a substrate to transfer lithographically defined patterns into underlying layers. Advanced etching techniques ensure high precision, accuracy, and reproducibility in PIC fabrication.

Wet Etching

Wet etching uses liquid chemical solutions to remove material from the substrate. It is often isotropic, meaning material is removed uniformly in all directions, which can lead to undercutting and less precise feature definition. However, selective etching allows targeted removal of specific materials without affecting others, making it useful for multilayer structures.

Dry Etching

Dry etching employs gases or plasmas to remove material and can be categorized into physical etching, chemical etching, and reactive ion etching (RIE).

- Physical Etching (Sputtering): High-energy ions bombard the substrate, physically removing material. This method provides high precision and anisotropy but lacks selectivity.

4.3 Advanced Lithography and Etching Methods: Detailed ...

- Chemical Etching: Reactive gases chemically interact with the substrate, forming volatile byproducts that are removed. It offers high selectivity but is generally isotropic.
- Reactive Ion Etching (RIE): A combination of physical and chemical etching, RIE generates a plasma of reactive gases, directing ions toward the substrate. This technique provides high precision, anisotropy, and selectivity, making it widely used in photonic device fabrication.

Deep Reactive Ion Etching (DRIE)

DRIE is a specialized form of RIE designed for creating deep, high-aspect-ratio structures. It often employs the Bosch process, alternating between etching and passivation steps to protect sidewalls, enabling highly anisotropic etching. DRIE is essential for fabricating deep trenches, vias, and waveguides used in advanced photonic devices.

Innovations in Lithography and Etching for Photonics

Directed Self-Assembly (DSA)

DSA uses block copolymers that naturally form nanoscale patterns through self-assembly. Lithographically defined guiding patterns direct the self-assembly process, enabling highly regular and precise nanoscale features. DSA is particularly useful in fabricating photonic crystals, waveguides, and other photonic structures that require extreme precision.

Atomic Layer Etching (ALE)

ALE is an advanced etching technique that removes material layer by layer with atomic-scale precision. By alternating between adsorption and desorption steps, ALE achieves controlled material removal with minimal substrate damage. It is used for high-precision nanostructures in photonic components, ensuring high aspect ratios and uniformity.

Applications of Advanced Lithography and Etching in Photonics
Waveguides

Precision lithography and etching create low loss waveguides with smooth surfaces and optimized geometries, ensuring efficient light propagation. These waveguides are essential for integrated photonic circuits used in telecommunications, data centers, and sensing applications.

Photonic Crystals

High-resolution lithography enables the fabrication of photonic crystals with precise periodic structures, creating photonic bandgaps that control light propagation. These structures are widely used in filters, resonators, and waveguides, enhancing photonic circuit performance.

Micro-optical Devices

Techniques like electron beam lithography (EBL) and two-photon polymerization (2PP) enable the fabrication of micro-lenses, gratings, and diffractive optical elements. These devices manipulate light with high precision, benefiting imaging, sensing, and communication technologies.

Quantum Photonic Devices

The precise patterning capabilities of advanced lithography and etching methods are critical for quantum photonic devices, including single-photon sources, quantum waveguides, and photonic qubits. These devices form the foundation of quantum computing and secure communication systems.

Future Directions
Lithography and etching techniques continue to evolve, with ongoing research aimed at further improving resolution, precision, and scalability.

- Next-Generation Lithography: Emerging techniques such as nanoimprint lithography (NIL) and directed self-assembly (DSA) are expected to play a significant role in future photonic device fabrication, offering high resolution with cost-effective production.
- Atomic-Scale Fabrication: Advances in atomic layer etching (ALE) and other atomic-scale fabrication techniques will enable the creation of next-generation photonic devices with superior precision and efficiency.
- Hybrid Fabrication: The combination of multiple lithography and etching techniques will allow for more complex and multifunctional photonic devices, leveraging the advantages of each method.

Advanced lithography and etching techniques are essential for the continued progress of photonic integrated circuits. Photolithography remains the dominant technique for mass production, while electron beam lithography (EBL), nanoimprint lithography (NIL), and extreme ultraviolet lithography (EUVL) enable higher resolution for specialized applications. Reactive ion etching (RIE), deep reactive ion etching (DRIE), and atomic layer etching (ALE) provide the precision required for nanostructured photonic devices. As fabrication techniques continue to advance, next-generation lithography and etching methods will drive innovation in photonics, unlocking new possibilities in telecommunications, quantum computing, sensing, and beyond.

Conclusion
Advanced lithography and etching methods are fundamental to the fabrication of photonic integrated circuits (PICs) and other micro- and nanoscale devices, as they enable the creation of intricate structures with sub-wavelength precision. Techniques like photolithography, electron beam lithography (EBL), and nanoimprint lithography (NIL) have

revolutionized the patterning of photonic structures, allowing for the precise control needed to build components such as waveguides, modulators, and resonators. Photolithography is the workhorse for large-scale production, offering rapid and cost-effective patterning, while EBL provides the high resolution required for complex, customized photonic devices at the nanoscale. Nanoimprint lithography, known for its ability to replicate fine patterns over large areas, has expanded the design possibilities for advanced photonics.

On the etching side, reactive ion etching (RIE) and its variations, like inductively coupled plasma (ICP) etching, allow for precise material removal, defining photonic components with sharp features and smooth sidewalls. These processes are critical in creating the structures that guide and manipulate light with minimal loss. Recent innovations, such as directed self-assembly (DSA) and atomic layer etching (ALE), are pushing the boundaries even further, offering unprecedented control at the atomic level. DSA uses molecular patterns to guide the formation of structures, providing a scalable solution for high-density integration, while ALE allows for atomic-scale precision in material removal, crucial for the development of next-generation photonic devices.

These advancements are enabling the fabrication of more compact, efficient, and powerful photonic devices, which are essential for applications in telecommunications, quantum computing, biophotonics, and beyond. As these technologies evolve, they will continue to drive innovation, enabling new applications in fields ranging from data centers to sensing technologies and medical imaging. This chapter delves into the principles and challenges associated with these advanced techniques, highlighting their role in shaping the future of photonics and integrated photonic systems.

Thermal Management in 3D PICs

5

- **Heat Dissipation Challenges**: Issues arising from increased packing density and hotspots.
- **Innovative Cooling Solutions**: Advanced materials and techniques for effective thermal management.
- **Materials with High Thermal Conductivity**: Selection of materials to improve heat dissipation.

5.1 Heat Dissipation Challenges: Issues Arising from Increased Packing Density and Hotspots in 3D Photonic Integrated Circuits

As Photonic Integrated Circuits (PICs) transition from 2D to 3D architectures, they offer substantial benefits in terms of integration density, allowing for more components to be packed into a smaller footprint. This leads to enhanced performance and functionality in a variety of applications, such as high-speed data transmission, advanced sensing, and quantum computing. However, the shift to 3D PICs also introduces significant thermal management challenges. As the packing density of active components increases, so does the potential for the formation of hotspots where heat accumulates, resulting in localized overheating. These thermal issues can degrade the performance of components, cause signal instability, and even lead to device failure if not effectively addressed. The vertical stacking of components in 3D architectures limits the surface area available for heat dissipation, making traditional cooling techniques, such as passive convection, less effective. Efficient thermal management is therefore crucial for maintaining system performance, reliability, and longevity. Potential solutions include the development of advanced cooling materials, such as high-thermal-conductivity layers, the incorporation of microfluidic cooling systems, and heat spreaders that can evenly distribute thermal loads across the chip. This chapter will explore the specific factors contributing to these thermal challenges and review current strategies and innovations designed to mitigate these issues, ensuring that 3D PICs can continue to operate efficiently under the increased power densities inherent to this architecture (Fig. 5.1).

Heat Dissipation in 3D PICs

Increased Packing Density

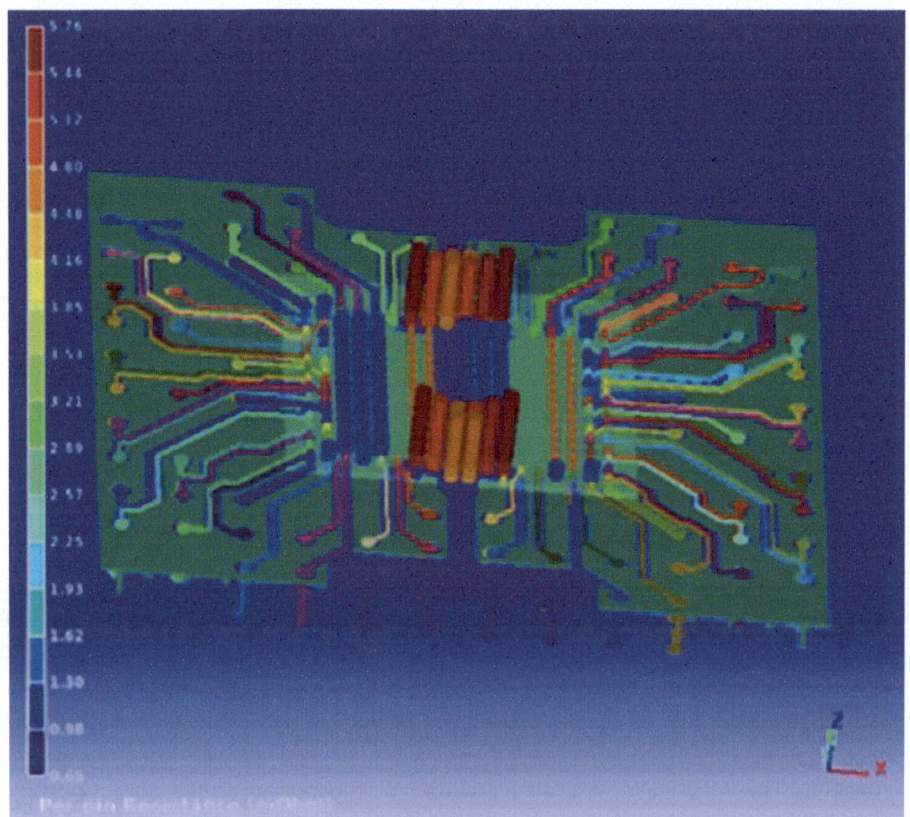

Fig. 5.1 Heat map visualizing hotspots in 3D PICs due to increased packing density. This is a resistance heatmap of a chip-package system with pin resolution The IR drop map and electromigration map can also be generated for power integrity and reliability sign-off of 3DIC system. https://semiwiki.com/eda/ansys-inc/314663-multiphysics-multivariate-analysis-an-imperative-for-todays-3d-ic-designs/

In 3D PICs, multiple layers of photonic and electronic components are stacked vertically, significantly increasing integration density compared to traditional 2D designs. While this approach enhances functionality and performance, it also results in greater heat generation within a smaller volume. Each layer generates heat during operation due to optical power absorption, electrical resistance, and electronic activity. The cumulative heat from multiple layers can lead to significant thermal challenges. Additionally, the vertical stacking of components limits the surface area available for heat dissipation, causing heat to accumulate within the device, elevating temperatures, and potentially leading to thermal runaway.

Hotspots

Hotspots are localized regions within a device where the temperature is significantly higher than the surrounding areas. In 3D PICs, hotspots can form due to uneven heat generation, poor thermal conductivity, and inadequate cooling. Certain components, such as lasers and modulators, generate more heat than others. When these high-power components are placed in close proximity within a 3D structure, they can create concentrated areas of excessive heat. Additionally, the materials used in 3D PICs, such as silicon and various dielectric layers, have different thermal conductivities. Discontinuities and interfaces between these materials can impede heat flow, exacerbating hotspot formation. Effective cooling mechanisms are also more difficult to implement in densely packed 3D structures. Traditional cooling methods, such as heat sinks and fans, may not be sufficient to address the thermal challenges posed by 3D integration.

Factors Contributing to Thermal Issues

Material Properties

The thermal properties of the materials used in 3D PICs play a significant role in heat dissipation. Materials with high thermal conductivity, such as silicon and diamond, efficiently transfer heat away from hot regions. However, many photonic materials, including certain dielectrics and polymers, have lower thermal conductivities, which hinder heat dissipation. Additionally, different materials expand at varying rates when exposed to heat. Mismatches in thermal expansion coefficients can cause mechanical stress and damage, further complicating thermal management and device reliability.

Device Design

The design and layout of 3D PICs significantly influence thermal performance. The placement of high-power components in close proximity can lead to localized heating and hotspots. To mitigate this, strategic placement and separation of these components are essential for effective thermal management. The configuration and number of layers also impact heat dissipation. Thicker layers and higher stacking densities can create barriers to heat flow, making it more challenging to maintain uniform temperatures across the device.

Operating Conditions

The operating conditions of 3D PICs, including power levels, duty cycles, and ambient temperature, also affect thermal behavior. Higher power levels result in increased heat generation, making thermal management more challenging as power density rises. Continuous operation at high power can cause thermal buildup, whereas implementing duty cycles with periods of lower power or rest can help manage temperatures. The ambient temperature of the surrounding environment also plays a role in heat dissipation. Higher external temperatures reduce the thermal gradient, making it harder to transfer heat away from the device, which can further degrade performance and longevity.

Effective thermal management strategies are critical for ensuring the stability and performance of 3D PICs, requiring a combination of material selection, device design optimization, and operational adjustments to mitigate heat-related challenges.

Potential Solutions for Thermal Management

Advanced Cooling Techniques

Implementing advanced cooling techniques is essential for managing heat in 3D PICs. One effective approach is microfluidic cooling, which integrates microfluidic channels within the 3D PIC structure to provide direct cooling to individual layers and components. Coolants, such as water or specialized fluids, efficiently transfer heat away from hot regions, preventing thermal buildup. Embedded microchannels are etched into the silicon substrate, allowing coolant to flow directly through the device, providing localized cooling and enhancing overall thermal management. Another method involves microjets, which direct coolant flow onto specific hotspots, delivering targeted cooling to high-power components (Fig. 5.2).

Another widely used method is thermoelectric cooling, which employs thermoelectric coolers (TECs) based on the Peltier effect to transfer heat from the device to a heat sink. TECs can be integrated into 3D PICs to provide active cooling, particularly in high-power applications. Peltier modules consist of thermoelectric materials that generate a cooling effect when an electric current passes through them. These modules can be strategically

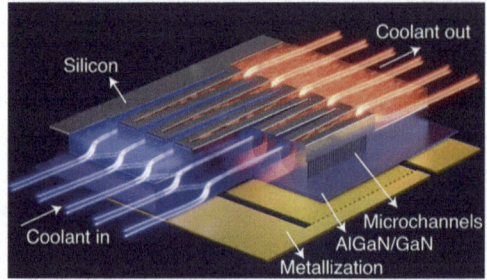

Fig. 5.2 Diagram of cooling solutions, such as microfluidic channels or heat spreaders, to manage heat dissipation. https://blog.darwin-microfluidics.com/glossary/microfluidic-cooling-microfluidics-explained/

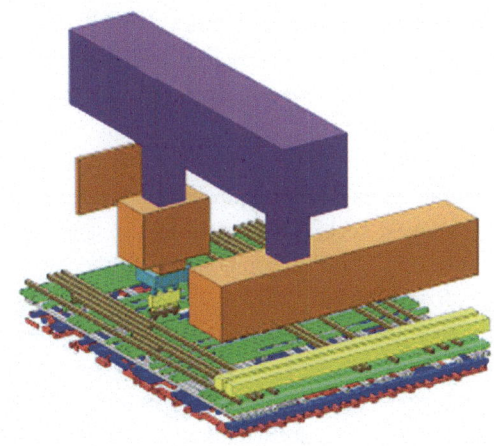

Fig. 5.3 Comparative illustration of materials with high thermal conductivity used in 3D PICs for effective heat management. Chrome-extension:// efaidnbmnnnibpcajpcglclefindmkaj/ https://poplab.stanford.edu/pdfs/Koroglu-HighTCinsulators3DICs-edl23.pdf

placed near heat-generating components, such as lasers and modulators, to help regulate temperatures and maintain device stability. By incorporating these advanced cooling strategies, 3D PICs can achieve improved thermal performance, ensuring reliability and efficiency in high-density photonic circuits.

Thermal Interface Materials (TIMs)

Thermal interface materials improve heat transfer between different layers and components in 3D PICs (Fig. 5.3).

High Thermal Conductivity TIMs play a crucial role in enhancing heat transfer between layers in 3D PICs. Materials such as thermal greases, pastes, and pads with high thermal conductivity can be applied at critical junctions to facilitate efficient heat dissipation. Among these, graphene-based TIMs offer exceptional thermal conductivity, making them highly effective in improving heat dissipation. Graphene and graphene oxide, in particular, are promising materials due to their superior thermal performance and compatibility with PIC fabrication processes.

Another effective thermal management solution is the use of Phase Change Materials (PCMs), which absorb and release thermal energy during phase transitions. This property allows PCMs to act as a thermal buffer, mitigating temperature fluctuations and preventing overheating in high-power operations. Encapsulated PCMs can be integrated within microchannels or strategically placed near hotspots to absorb excess heat during peak power loads and release it when the system operates at lower power levels. This dynamic heat management approach helps maintain stable operating temperatures, improving the overall reliability and efficiency of 3D PICs.

Thermal Design Optimization

Optimizing the thermal design of 3D PICs requires strategic approaches to enhance heat dissipation and minimize hotspots. Component placement is a critical factor, as positioning high-power components in a way that evenly distributes heat generation can significantly improve overall thermal management. Placing heat-sensitive elements away from hotspots and high-power regions helps maintain stable operation. Thermal isolation techniques, such as incorporating air gaps or low-conductivity materials between high-power and sensitive components, further prevent thermal crosstalk and the formation of localized hotspots.

Thermal via arrays play a crucial role in facilitating vertical heat transfer within 3D structures. These vias serve as highly conductive pathways, allowing heat to be efficiently conducted away from high-temperature regions. Copper vias, due to their excellent thermal conductivity, are widely used for this purpose. Strategically placing arrays of copper vias within the structure ensures effective heat flow management, reducing the risk of thermal buildup.

Thermal simulation and modeling provide valuable insights into the thermal behavior of 3D PICs, allowing engineers to identify potential hotspots and optimize layouts for improved heat dissipation. Finite Element Analysis (FEA) is a particularly powerful tool for simulating heat distribution across different layers, offering detailed data that can inform design refinements to enhance overall thermal performance.

Advanced Materials

Developing and integrating advanced materials with superior thermal properties is a key strategy for improving heat dissipation in 3D PICs. Diamond and diamond-like carbon (DLC) offer exceptional thermal conductivity and can be used as substrates or heat spreaders. Diamond-coated substrates provide enhanced thermal management while maintaining compatibility with standard fabrication processes.

Another promising material is carbon nanotubes (CNTs), which exhibit high thermal conductivity and mechanical strength, making them well-suited for heat dissipation applications in 3D PICs. CNT thermal interfaces, when incorporated between layers or used as fillers in thermal interface materials (TIMs), significantly enhance heat transfer efficiency, reducing the overall thermal resistance of the device.

Case Studies and Applications

High-Power Lasers

High-power lasers in 3D PICs generate substantial heat, posing a significant thermal management challenge. One effective solution is microfluidic cooling, where microchannels are integrated directly beneath laser components to facilitate efficient heat dissipation. This approach prevents thermal degradation and ensures stable laser operation.

Optical Interconnects in Data Centers

Data centers rely on high-performance optical interconnects, which require efficient thermal management to maintain signal integrity and low latency. Thermoelectric cooling is a viable solution, as integrating thermoelectric coolers (TECs) with 3D PIC-based optical interconnects helps regulate heat generation, preventing signal degradation and improving overall system efficiency.

Quantum Photonic Devices

Quantum photonic devices are highly sensitive to temperature fluctuations, requiring precise thermal control to maintain stability. Graphene-based TIMs are particularly beneficial in quantum photonic applications, as their excellent thermal conductivity ensures stable operating temperatures. This stability is crucial for reliable performance in quantum computing and secure communication systems.

Future Directions

Ongoing research in thermal management for 3D PICs is focused on several key advancements:

- Nanostructured Materials: The development of novel nanostructured materials with tailored thermal properties is expected to significantly enhance heat dissipation capabilities.
- Integrated Cooling Systems: Researchers are exploring hybrid cooling solutions that combine multiple techniques, such as microfluidic cooling, thermoelectric cooling, and advanced TIMs, to achieve optimal thermal performance.
- AI-Driven Thermal Design: The use of artificial intelligence and machine learning to optimize thermal design is emerging as a promising approach. AI-driven simulations can predict thermal behavior in complex 3D PICs, enabling automated design adjustments for enhanced heat management.

As these technologies continue to advance, the integration of innovative thermal management strategies will play a crucial role in unlocking the full potential of 3D PICs across a wide range of applications.

Conclusion

Heat dissipation challenges are a critical concern in the development of 3D Photonic Integrated Circuits. Increased packing density and the formation of hotspots can significantly impact the performance and reliability of these advanced devices. Addressing these challenges requires a multi-faceted approach, including advanced cooling techniques, thermal interface materials, thermal design optimization, and the development of advanced materials. By leveraging these strategies, it is possible to enhance the thermal management of 3D PICs, ensuring their continued evolution and success in various high-performance applications.

5.2 Innovative Cooling Solutions: Advanced Materials and Techniques for Effective Thermal Management in 3D Photonics

As 3D Photonic Integrated Circuits (PICs) become more complex and densely packed, managing heat dissipation effectively is critical to preserving their performance, reliability, and longevity. The increased integration density in 3D architectures results in higher power densities and localized thermal hotspots, which can degrade the efficiency and functionality of photonic devices. Traditional passive cooling methods, such as heat sinks or simple convection, are insufficient to handle the thermal loads in these advanced systems, making innovative cooling solutions imperative. This chapter delves into several cutting-edge approaches for addressing these thermal management challenges. Advanced materials like high-thermal-conductivity layers, including graphene and diamond, are being explored for their exceptional heat dissipation properties, allowing more efficient transfer of heat away from critical components. Additionally, the integration of microfluidic cooling systems within the chip architecture provides an active approach to manage localized hotspots by circulating coolant through tiny channels embedded in the photonic structure. Other strategies, such as heat spreaders that distribute thermal loads more evenly across the chip and thermal vias that guide heat through the vertical layers of the 3D stack, are gaining traction as potential solutions. Together, these innovations not only mitigate the risk of thermal overload but also ensure that 3D PICs can achieve their full potential in applications such as telecommunications, data centers, and quantum computing. This chapter will explore these advanced cooling techniques and their role in driving the next generation of 3D PIC technologies.

Advanced Materials for Thermal Management
High Thermal Conductivity Materials

Materials with high thermal conductivity are essential for efficient heat dissipation in 3D PICs. These materials help rapidly transfer heat away from hotspots, minimizing temperature gradients and ensuring uniform thermal distribution. Diamond and Diamond-Like Carbon (DLC) exhibit exceptional thermal conductivity, making them ideal for heat spreaders and substrates in 3D PICs. Diamond-coated substrates and DLC layers can enhance heat dissipation in high-power components such as lasers and modulators. Graphene, known for its extraordinary thermal conductivity, mechanical strength, and flexibility, is another versatile material for thermal management. Graphene-based thermal interface materials (TIMs) and heat spreaders can significantly improve heat transfer between layers and components in 3D PICs. Similarly, Carbon Nanotubes (CNTs) offer high thermal conductivity and excellent mechanical properties, providing efficient pathways for heat dissipation. CNT arrays can be integrated as thermal vias or fillers in TIMs to enhance vertical heat transfer in 3D PIC structures.

Phase Change Materials (PCMs)

Phase Change Materials (PCMs) absorb and release thermal energy during phase transitions, providing a buffering effect that helps manage temperature fluctuations. Encapsulated PCMs can be integrated within microchannels or embedded in TIMs to absorb excess heat during high-power operation and release it during lower power periods. These materials are particularly useful in regions with significant thermal cycling, such as near high-power lasers and modulators.

Hybrid Materials

Hybrid materials combine the advantages of different thermal management materials to enhance overall performance. Metal-Organic Frameworks (MOFs), which are porous materials engineered for high thermal conductivity, can be used as thermal interface materials or integrated into substrates to improve heat dissipation and mechanical stability. Composite materials that blend graphene, CNTs, and high thermal conductivity polymers offer tailored thermal properties and structural integrity, making them suitable for TIMs and substrates to enhance thermal performance and durability.

Innovative Cooling Techniques

Microfluidic Cooling involves the use of microchannels to circulate coolant fluids, providing efficient and localized cooling. Embedded microchannels can be etched into the silicon substrate or other layers of the 3D PIC, allowing coolant to flow directly through the device. This approach provides localized cooling for high-power components such as lasers and modulators, reducing the risk of thermal hotspots. Microjets, which direct coolant flow onto specific regions or components, offer targeted cooling to ensure uniform temperature distribution and prevent thermal buildup.

Thermoelectric Cooling utilizes thermoelectric coolers (TECs) based on the Peltier effect to transfer heat from the device to a heat sink, providing active cooling. Peltier modules consist of thermoelectric materials that generate a cooling effect when an electric current passes through them. These modules can be integrated near high-power components to manage their temperatures effectively, enhancing performance and reliability.

Advanced Thermal Interface Materials (TIMs)

TIMs improve heat transfer between different layers and components in 3D PICs, ensuring efficient thermal management. Graphene-based TIMs, due to their high thermal conductivity, enhance heat transfer between layers and can be applied between high-power components and heat spreaders. Additionally, nanoparticle-enhanced TIMs, incorporating materials such as silver or copper, improve thermal conductivity and overall performance. These materials are especially useful in regions with high thermal resistance, enhancing heat dissipation and reducing thermal gradients.

Thermal Design Optimization

Optimizing the thermal design of 3D PICs involves strategic placement and configuration of components to enhance heat dissipation and minimize hotspots. Component placement is crucial, as positioning high-power components strategically ensures even heat distribution across the device. Thermal isolation techniques, such as air gaps or low-conductivity materials, help prevent thermal crosstalk. The introduction of thermal via arrays—highly conductive pathways within the 3D structure—enhances vertical heat transfer. Arrays of copper or CNT thermal vias can be strategically placed to manage heat flow effectively, connecting different layers and improving thermal performance.

Advanced Simulation and Modeling

Using advanced simulation tools to model the thermal behavior of 3D PICs during the design phase helps identify potential hotspots and optimize the layout for better thermal performance. Finite Element Analysis (FEA) provides detailed insights into heat distribution and highlights areas where thermal management can be improved. FEA can simulate the thermal behavior of different materials and cooling techniques, guiding the design of more efficient thermal management systems. Additionally, machine learning and AI algorithms can analyze thermal data and predict thermal behavior, optimizing the design and placement of cooling systems. AI-driven thermal design enhances the efficiency of cooling solutions, ensuring optimal performance under varying operating conditions.

Case Studies and Applications

High-Power Lasers generate substantial heat, posing a significant thermal management challenge in 3D PICs. Implementing microfluidic cooling directly beneath laser components helps dissipate heat efficiently, preventing thermal degradation and ensuring stable operation.

Optical Interconnects in Data Centers require high-performance thermal management. Integrating thermoelectric coolers with 3D PIC-based optical interconnects helps manage heat generation, maintaining signal integrity and reducing latency.

Quantum Photonic Devices are highly sensitive to temperature fluctuations, necessitating precise thermal management. Using graphene-based TIMs in quantum photonic devices enhances thermal stability, ensuring reliable performance in quantum computing and communication applications.

Biosensors and Medical Diagnostics often operate in environments where precise thermal management is critical for accurate measurements. Integrating microfluidic cooling channels within biosensors ensures consistent operating temperatures, enhancing sensitivity and reliability.

Future Directions

Ongoing research and development in thermal management for 3D PICs focus on several key areas. The development of nanostructured materials with tailored thermal properties will further enhance heat dissipation. The design of integrated cooling systems that combine multiple cooling techniques will optimize thermal management. Additionally, AI-driven thermal design using artificial intelligence and machine learning will refine the optimization of thermal designs and predict thermal behavior in complex 3D PICs. These advancements will play a critical role in the continued evolution of photonic integrated circuits, enabling higher performance, greater reliability, and expanded applications across various industries.

Conclusion

Innovative cooling solutions are critical for managing heat dissipation in 3D Photonic Integrated Circuits. Advanced materials, such as diamond, graphene, and carbon nanotubes, offer superior thermal conductivity and mechanical properties, enhancing heat transfer and thermal stability. Novel cooling techniques, including microfluidic cooling and thermoelectric coolers, provide efficient and localized thermal management, preventing hotspots and maintaining uniform temperature distribution. By leveraging these advanced materials and techniques, it is possible to overcome the thermal challenges associated with 3D PICs, ensuring their continued evolution and success in various high-performance applications.

5.3 Materials with High Thermal Conductivity: Selection of Materials to Improve Heat Dissipation in 3D Photonics

Effective thermal management is a crucial consideration in the design and operation of 3D Photonic Integrated Circuits (PICs). As these devices become increasingly complex, with higher integration density and a greater number of active photonic and electronic components packed into smaller volumes, the challenge of managing the heat generated by these components intensifies. Without proper heat dissipation, thermal hotspots can form, leading to performance degradation, signal loss, and potential device failure. To address these issues, materials with high thermal conductivity play a pivotal role in efficiently conducting heat away from critical regions. Materials such as diamond, graphene, and silicon carbide (SiC) have emerged as promising solutions due to their exceptional thermal properties. Diamond, for instance, has one of the highest thermal conductivities of any material, making it ideal for use in heat spreaders or as a thermal interface material. Graphene, with its high in-plane thermal conductivity, is being explored for its ability to dissipate heat in compact devices. Meanwhile, silicon carbide offers a balance between thermal conductivity and semiconductor properties, making it useful for both photonic and electronic applications. This chapter will explore the properties and advantages of these and other advanced materials, along with their applications in improving heat dissipation in 3D PICs, ensuring that these cutting-edge systems can operate at peak performance even under high-power conditions. By addressing these thermal challenges, designers can significantly enhance the reliability and lifespan of 3D photonic systems across applications in telecommunications, data centers, and quantum computing.

Importance of Thermal Management in 3D PICs

The performance and longevity of 3D PICs depend significantly on efficient thermal management. Poor heat dissipation can lead to increased operating temperatures, thermal crosstalk between components, and potential thermal runaway, all of which degrade device performance and reliability.

Challenges in 3D PICs

The vertical stacking of multiple layers in 3D PICs results in higher power densities and increased heat generation. However, the compact design limits the available surface area for heat dissipation, making effective cooling more challenging. Additionally, localized regions with high power density can create hotspots, leading to uneven temperature distribution and potential failure points.

High Thermal Conductivity Materials

Selecting materials with high thermal conductivity is essential for improving heat dissipation in 3D PICs. These materials efficiently transfer heat away from hotspots, maintaining uniform temperature distribution across the device.

Diamond

Diamond is one of the best materials for thermal management due to its exceptional thermal conductivity, which can reach up to 2200 W/m K. In addition to its superior heat conduction properties, diamond is extremely hard and has a high Young's modulus, providing structural integrity. Diamond can be used as a substrate or heat spreader in high-power photonic components, such as lasers and modulators. Additionally, diamond coatings can enhance the thermal performance of silicon substrates in 3D PICs.

Graphene

Graphene, a single layer of carbon atoms arranged in a hexagonal lattice, exhibits outstanding thermal properties. With a thermal conductivity of approximately 5000 W/m K, graphene is an excellent material for heat dissipation. It is also both strong and flexible, allowing it to conform to various shapes and surfaces. Graphene-based thermal interface materials (TIMs) and heat spreaders can enhance heat transfer between layers and components in 3D PICs. It can also be integrated into substrates to improve overall thermal management.

Carbon Nanotubes (CNTs)

Carbon nanotubes are cylindrical structures composed of carbon atoms with remarkable thermal and mechanical properties. Their thermal conductivity ranges from 3000 to 6000 W/m K, depending on their structure and alignment. CNTs are incredibly strong and can withstand high temperatures without degradation. They can be used as thermal vias to enhance vertical heat transfer in 3D PICs and can also be incorporated into TIMs to improve heat dissipation between interfaces.

Silicon Carbide (SiC)

Silicon carbide is a compound semiconductor material known for its high thermal conductivity and mechanical robustness. With a thermal conductivity of about 490 W/m K, SiC significantly outperforms silicon in heat dissipation. It can operate at higher temperatures without losing its thermal and mechanical properties, making it ideal for use as a substrate or heat spreader in high-power photonic devices. SiC is also suitable for environments with high thermal loads and extreme conditions.

Aluminum Nitride (AlN)

Aluminum nitride is a ceramic material with excellent thermal conductivity and electrical insulation properties. With a thermal conductivity of about 285 W/m K, it serves as an effective heat dissipator while also providing electrical isolation. AlN can be used as a substrate for photonic devices that require both high thermal conductivity and electrical insulation. It is also commonly incorporated into TIMs to enhance heat transfer while maintaining electrical isolation.

Applications of High Thermal Conductivity Materials

Substrates and Heat Spreaders

Using high thermal conductivity materials as substrates and heat spreaders is essential for effective heat dissipation in 3D PICs. Diamond substrates provide superior heat dissipation for high-power devices, reducing thermal gradients and improving performance. Graphene heat spreaders enhance thermal management by distributing heat evenly across the device, preventing hotspots. SiC and AlN substrates offer both high thermal conductivity and stability, making them suitable for high-temperature and high-power applications.

Thermal Interface Materials (TIMs)

TIMs improve heat transfer between different layers and components in 3D PICs, ensuring efficient thermal management. Graphene-based TIMs provide excellent thermal conductivity, reducing thermal resistance between interfaces. CNT-enhanced TIMs offer high thermal conductivity and mechanical strength, improving heat dissipation in high-power applications. AlN TIMs combine good thermal conductivity with electrical insulation, making them ideal for applications where both properties are required.

Thermal Vias

Thermal vias are conductive pathways that enhance vertical heat transfer in 3D PICs, connecting different layers and improving overall thermal management. Copper is commonly used for thermal vias due to its high thermal conductivity, with arrays of copper vias strategically placed to manage heat flow effectively. CNT thermal vias offer even higher thermal conductivity and mechanical strength compared to traditional materials, making them suitable for advanced thermal via applications.

Case Studies and Applications

High-Power Lasers

High-power lasers generate substantial heat, requiring efficient thermal management solutions. Diamond-coated substrates significantly enhance heat dissipation, preventing thermal degradation and ensuring stable operation. Graphene TIMs improve heat transfer between the laser and the heat sink, maintaining optimal operating temperatures and performance.

Optical Interconnects in Data Centers

Data centers require high-performance optical interconnects with efficient thermal management to handle high data throughput. Integrating graphene heat spreaders in optical interconnects enhances thermal management, reducing latency and improving signal integrity. CNT thermal vias provide efficient vertical heat transfer, ensuring uniform temperature distribution across the 3D PIC.

Quantum Photonic Devices

Quantum photonic devices are highly sensitive to temperature fluctuations, necessitating precise thermal management. Using graphene-based TIMs in quantum photonic devices enhances thermal stability, ensuring reliable performance in quantum computing and communication applications.

Biosensors and Medical Diagnostics

Biosensors and medical diagnostic devices require precise thermal management for accurate measurements. Integrating microfluidic cooling channels with high thermal conductivity materials ensures consistent operating temperatures, enhancing sensitivity and reliability. Graphene and CNT TIMs improve heat dissipation in biosensors, maintaining optimal conditions for accurate detection and measurement.

Future Directions

Ongoing research and development in high thermal conductivity materials focus on several key areas. The development of nanostructured materials with tailored thermal properties

5.3 Materials with High Thermal Conductivity: Selection ...

aims to enhance heat dissipation in 3D PICs. Creating composite materials that combine the advantages of different thermal management materials will offer tailored thermal properties and structural integrity. AI-driven material design using artificial intelligence and machine learning will help optimize high thermal conductivity materials for specific applications in 3D photonics, enabling more efficient and reliable devices for the future.

Materials with high thermal conductivity are pivotal in managing the heat generated within 3D Photonic Integrated Circuits (PICs). As PICs integrate more photonic and electronic components within compact 3D architectures, effective thermal management becomes essential to maintaining their performance, reliability, and longevity. The primary challenge lies in dissipating heat from densely packed active components, which can otherwise lead to overheating, thermal hotspots, and eventual system failure. Advanced materials with superior thermal properties, such as diamond, graphene, carbon nanotubes (CNTs), silicon carbide (SiC), and aluminum nitride (AlN), offer a range of solutions for overcoming these challenges, enabling the continued advancement of 3D PIC technology.

1. **Diamond:**

Diamond is often regarded as the best material for heat dissipation due to its extremely high thermal conductivity (around 2000 W/m K), which is several times greater than that of silicon. Synthetic diamond films can be used as heat spreaders in 3D PICs, conducting heat away from active photonic components with minimal resistance. Diamond is particularly useful in high-power photonic devices, such as lasers and modulators, where excessive heat could degrade performance. Moreover, diamond's thermal conductivity remains high even at elevated temperatures, making it suitable for extreme operating conditions. In some cases, diamond-like carbon (DLC) films are also employed as a cost-effective alternative with excellent thermal properties.

2. **Graphene:**

Graphene, a two-dimensional material composed of a single layer of carbon atoms arranged in a honeycomb lattice, exhibits remarkable in-plane thermal conductivity, typically in the range of 3000–5000 W/m K. Its unique properties make it an excellent candidate for integrating into 3D PICs as a thermal interface material or for use in flexible photonic components. Due to its high mechanical strength and excellent flexibility, graphene can be embedded within multilayer photonic structures to dissipate heat while maintaining the structural integrity of the device. Moreover, its ability to be layered and patterned with other materials makes it an ideal candidate for hybrid cooling systems in compact photonic circuits.

3. **Carbon Nanotubes (CNTs):**

Carbon nanotubes are another promising material for thermal management in 3D PICs. CNTs have exceptional thermal conductivity along their axis, ranging from 3000 to 6000 W/m K, and can be used to create thermal interconnects that efficiently transfer heat between different layers of a 3D PIC. They also offer high surface area and flexibility, allowing them to be incorporated into tight spaces between photonic components. In addition to their thermal properties, CNTs are lightweight and can be used in multifunctional roles where both electrical conductivity and heat dissipation are required, such as in integrated optoelectronic devices.

4. **Silicon Carbide (SiC):**

Silicon carbide is known for its combination of high thermal conductivity (about 120–270 W/m K) and semiconductor properties, making it an ideal material for high-power electronics and photonic devices. SiC is often used in optoelectronic components, such as LEDs and laser diodes, where thermal management is critical to performance and efficiency. In 3D PICs, SiC can serve as both a heat-dissipating substrate and an active material, enabling devices to operate at higher temperatures without degradation. SiC's high thermal stability also makes it ideal for applications in harsh environments, such as those encountered in aerospace and defense.

5. **Aluminum Nitride (AlN):**

Aluminum nitride offers an excellent balance between thermal conductivity (170–320 W/m K) and electrical insulation, making it a valuable material for use in photonic integrated circuits where both electrical and thermal management are required. AlN is commonly used as a substrate material or thermal interface layer, ensuring efficient heat removal from high-power photonic components while maintaining electrical isolation. Its compatibility with standard semiconductor processing techniques further enhances its applicability in 3D PICs. Additionally, AlN is increasingly used in optoelectronics, such as UV LEDs and high-frequency devices, where both heat and electrical isolation are critical.

Future Prospects and Research:
As research and development in the field of 3D photonics continue to progress, the role of these high thermal conductivity materials will become even more critical. Emerging technologies, such as heterogeneous integration, which combines different materials and components on a single chip, will further benefit from the thermal management properties of diamond, graphene, CNTs, SiC, and AlN. Additionally, nano-engineered materials and composite structures that incorporate these high thermal conductivity materials are

being developed to provide tailored thermal properties for specific applications. This will enable the design of next-generation 3D PICs that can operate at higher powers, with greater efficiency and reliability, across industries such as telecommunications, data centers, automotive systems, and quantum computing.

Conclusion:

In conclusion, materials with high thermal conductivity, such as diamond, graphene, carbon nanotubes, silicon carbide, and aluminum nitride, are essential to addressing the thermal challenges faced by 3D Photonic Integrated Circuits. By leveraging the exceptional thermal properties of these materials, designers and engineers can ensure efficient heat dissipation, enabling the continued advancement of 3D PICs in high-performance applications. As thermal management remains a critical issue in the ongoing evolution of 3D photonics, the role of these advanced materials will continue to be at the forefront of innovation, driving the development of more reliable, efficient, and powerful photonic systems.

6. Alignment and Packaging of 3D PICs

- **Precision Alignment Techniques**: Methods to maintain optical signal integrity and ensure accurate positioning.
- **Packaging Technologies**: Strategies for robust and efficient packaging of 3D PICs.
- **Inter-layer Optical Interconnects**: Solutions for vertical optical connections with low loss and high misalignment tolerance.

6.1 3D Photonics Precision Alignment Techniques: Methods to Maintain Optical Signal Integrity and Ensure Accurate Positioning

The precision alignment of components in 3D Photonic Integrated Circuits (PICs) is crucial for maintaining optical signal integrity and ensuring that each element is positioned accurately to achieve optimal device performance. As photonic devices advance toward higher complexity and miniaturization, precise alignment becomes increasingly challenging, as even microscopic misalignments can lead to significant signal losses, increased crosstalk, and degraded device efficiency. Alignment techniques in 3D PICs involve sophisticated methods such as active alignment, where optical signals are monitored in real-time during assembly to ensure peak performance, and passive alignment, which relies on pre-defined mechanical structures or alignment marks that aid in accurate positioning without continuous monitoring. Additionally, techniques like wafer bonding and self-aligned lithography enable precise stacking and integration of photonic layers, which are essential for building multi-layered 3D structures with high-density interconnects.

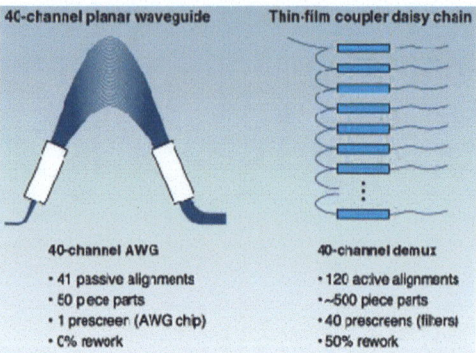

Fig. 6.1 Illustration of precision alignment techniques showing the alignment of waveguides across layers. https://www.laserfocusworld.com/software-accessories/positioning-support-accessories/article/16556682/silica-on-silicon-remains-a-staple-for-making-waveguides

This chapter examines these and other precision alignment methods in detail, highlighting innovations that allow for enhanced optical signal fidelity and positioning accuracy, which are essential for the effective scaling and functionality of 3D photonic devices (Fig. 6.1).

Importance of Precision Alignment in 3D Photonics

Accurate alignment is essential for the efficient coupling of light between photonic components, such as waveguides, lasers, modulators, and detectors. Misalignment can lead to significant optical losses, signal degradation, and reduced device performance. Ensuring precise alignment is crucial in 3D photonic integrated circuits (PICs), where complex structures and vertical stacking introduce additional alignment challenges.

Challenges in 3D Photonics Alignment

The intricate and compact structures of 3D PICs make alignment more challenging than in 2D systems. Misalignment between layers in vertically stacked structures can disrupt optical pathways, leading to signal loss. Additionally, different materials expand at different rates when heated, causing potential misalignment due to thermal expansion.

Passive Alignment Techniques

Passive alignment relies on physical features and precision fabrication to achieve accurate positioning of photonic components without active adjustment during assembly.

Lithographic Alignment

Lithographic alignment uses precise photolithography processes to define the positions of photonic components with high accuracy. During photolithography, masks with predefined patterns are aligned with the substrate to ensure accurate placement of features. Advanced mask aligners use optical and mechanical systems to achieve sub-micron alignment precision. Step-and-repeat lithography involves sequential exposure of the substrate using a stepper, aligning each exposure to the previous one with high precision, ensuring consistent alignment across the entire wafer.

Self-alignment Techniques

Self-alignment techniques use the natural tendencies of materials and structures to align themselves during fabrication and assembly. Capillary force alignment leverages the surface tension of liquids, such as solder or adhesive droplets, to pull components into precise alignment during bonding. Directed self-assembly (DSA) uses block copolymers that self-organize into well-defined nanostructures, directed by lithographically defined guiding patterns to achieve precise alignment.

Mechanical Alignment Structures

Mechanical alignment structures are built into the substrate or components to facilitate passive alignment during assembly. Precision-machined alignment pins and holes ensure correct positioning by fitting components into predefined slots, minimizing misalignment. V-grooves and ridges etched into the substrate guide optical fibers or waveguides into precise positions, a technique commonly used in fiber-optic alignment.

Active Alignment Techniques

Active alignment involves real-time feedback and adjustment during assembly to achieve precise positioning of photonic components. This method is often used for high-precision applications where passive alignment is insufficient.

Optical Feedback Alignment

Optical feedback alignment uses light signals to monitor and adjust component positions during assembly. In power monitoring, light is coupled into the photonic circuit, and the output power is measured. Adjustments are made to maximize optical power, indicating optimal alignment. Interferometric alignment, which measures phase and amplitude variations of light signals, achieves sub-wavelength precision by analyzing interference patterns and making precise adjustments accordingly.

Robotic Assembly

Robotic assembly systems equipped with high-resolution cameras and precision actuators can achieve accurate alignment through automated processes. Pick-and-place robots use vacuum grippers or mechanical arms to pick up and position components, adjusting placement in real-time with vision feedback. Nano-manipulators provide ultra-precise control over component positioning, enabling alignment at the nanometer scale, which is essential for highly integrated 3D PICs.

Hybrid Alignment Techniques

Hybrid alignment techniques combine passive and active alignment methods to achieve high precision and reliability in 3D photonics.

Pre-alignment and Fine Adjustment

Pre-alignment involves using passive techniques to achieve coarse alignment, followed by active techniques for fine adjustment. Mechanical structures, such as pins and V-grooves, provide an initial alignment, after which optical feedback adjustments refine positioning to achieve precise alignment.

Multi-stage Alignment

Multi-stage alignment sequentially aligns different components or layers of a 3D PIC using a combination of passive and active techniques. Each layer of a 3D PIC is aligned and bonded one at a time, using optical feedback and mechanical alignment structures to ensure accuracy. Integrated feedback systems provide real-time data for each stage of alignment, ensuring consistent precision throughout the entire 3D structure.

Case Studies and Applications

Silicon Photonics

Silicon photonics requires precise alignment of waveguides, modulators, and detectors to achieve efficient light coupling and signal integrity. Passive alignment techniques, such as V-grooves and alignment pins, are commonly used for fiber-optic alignment in silicon photonic circuits. Active optical feedback alignment is employed for precise coupling of laser light into silicon waveguides, maximizing optical power and minimizing losses.

Data Centers and Optical Interconnects

High-speed optical interconnects in data centers require precise alignment to ensure low-loss signal transmission. Robotic assembly systems with vision feedback align optical fibers and connectors with high precision, ensuring reliable performance in data center networks. Hybrid alignment methods are used in optical switches, where passive pre-alignment is combined with active fine adjustments to achieve optimal performance.

Quantum Photonics

Quantum photonic devices are highly sensitive to alignment, requiring ultra-precise positioning to maintain quantum coherence and signal integrity. Nano-manipulators are used to align single-photon sources and detectors with nanometer precision, ensuring reliable operation of quantum photonic circuits. Interferometric alignment techniques provide the precision needed to align components in quantum photonic devices while maintaining the integrity of quantum states.

Future Directions

The future of precision alignment in 3D photonics involves the development of more advanced techniques and technologies to enhance accuracy and efficiency. AI-driven alignment systems utilizing machine learning algorithms can analyze alignment data and optimize the alignment process in real-time, improving accuracy and reducing assembly time. The development of advanced metrology tools, such as high-resolution imaging and spectroscopy, will provide better feedback for alignment processes, enabling sub-nanometer precision. Integrating alignment systems directly into the fabrication process will streamline assembly and improve consistency, ensuring high precision and reliability in 3D PICs.

Conclusion

Precision alignment techniques are essential for maintaining optical signal integrity and ensuring accurate positioning in 3D photonic integrated circuits. Passive alignment methods, such as lithographic alignment, self-alignment, and mechanical alignment structures, provide reliable positioning for many applications. Active alignment techniques, including optical feedback and robotic assembly, offer high precision for more demanding requirements. Hybrid alignment approaches combine the strengths of both passive and active methods to achieve optimal results. By leveraging these advanced alignment techniques, researchers and engineers can overcome the challenges of 3D photonics, ensuring the performance and reliability of next-generation photonic devices.

6.2 Packaging Technologies: Strategies for Robust and Efficient Packaging of 3D PICs

The advancement of 3D Photonic Integrated Circuits (PICs) demands innovative packaging technologies to support their complex architectures and ensure consistent performance and reliability over time. Packaging for 3D PICs goes beyond simple protection, as it must address multiple critical factors, including environmental shielding, thermal management, signal integrity, and mechanical stability. Due to the high density of components and vertical stacking in 3D PICs, thermal management is essential to prevent overheating, which can degrade performance and shorten device lifespan. This requires packaging solutions that integrate heat spreaders, thermal interfaces, or microfluidic cooling systems to efficiently dissipate heat. Additionally, maintaining signal integrity in tightly packed 3D PICs is challenging due to potential interference between layers; advanced materials and alignment techniques in packaging help minimize signal loss and crosstalk. The packaging also contributes to mechanical robustness by protecting fragile photonic components from external stresses and environmental factors like humidity, dust, and temperature fluctuations. This chapter delves into a range of packaging strategies used in 3D photonics, highlighting cutting-edge techniques such as wafer-level packaging, flip-chip bonding, and

Fig. 6.2 Schematic of packaging technologies for 3D PICs, showing protective layers and interfaces. https://www.mdpi.com/2076-3417/6/12/426

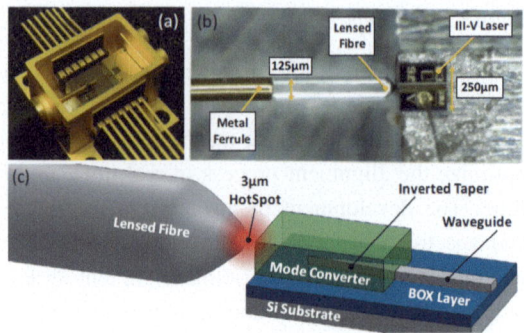

hermetic sealing, all of which enhance the durability, performance, and efficiency of 3D PICs, paving the way for their adoption in demanding applications such as data centers, telecommunications, and advanced sensing systems (Fig. 6.2).

Importance of Packaging in 3D PICs

Packaging plays a critical role in ensuring the performance, reliability, and longevity of 3D Photonic Integrated Circuits (PICs). It serves multiple essential functions, including protecting photonic components from physical damage, dust, moisture, and other environmental factors. Effective packaging facilitates heat dissipation to maintain optimal operating temperatures and prevent thermal degradation. Additionally, it ensures the efficient transmission of optical and electrical signals with minimal loss and interference while providing structural support to maintain the integrity of the 3D architecture.

Packaging Strategies for 3D PICs

Hermetic Packaging

Hermetic packaging involves sealing the PIC in an airtight enclosure to protect it from moisture, dust, and other environmental contaminants. Common materials for hermetic packaging include metals such as Kovar and ceramics, both of which offer excellent barrier properties. Sealing methods such as welding, soldering, or glass-frit bonding are employed to create a robust hermetic enclosure. This type of packaging is ideal for environments with harsh conditions, such as aerospace, military, and industrial applications.

Flip-Chip Bonding

Flip-chip bonding mounts the PIC upside-down on a substrate, allowing for direct electrical and thermal contact. Metal bumps, typically made of solder or gold, are deposited on the chip's bonding pads to provide electrical connections and mechanical support. An underfill material is applied between the chip and substrate to enhance mechanical stability and thermal performance. Flip-chip bonding offers shorter interconnect lengths, reducing electrical resistance and improving thermal dissipation, making it widely used in high-performance computing and telecommunications.

Wire Bonding

Wire bonding is a traditional packaging technique where thin wires connect the chip's bonding pads to external leads. Gold and aluminum wires are commonly used due to their excellent electrical conductivity and bondability. Techniques such as thermocompression bonding, ultrasonic bonding, and thermosonic bonding are employed to create reliable connections. Wire bonding is versatile and cost-effective, making it suitable for a wide range of applications, though it may not be ideal for high-density interconnects due to potential signal delay and crosstalk.

Through-Silicon Vias (TSVs)

TSVs are vertical electrical connections that pass through the silicon substrate, enabling direct interlayer communication in 3D PICs. These are fabricated by etching holes through the silicon wafer and filling them with a conductive material such as copper. In addition to providing electrical interconnects, TSVs function as thermal vias, facilitating heat dissipation from inner layers to heat sinks. They are critical for high-performance computing and advanced photonic applications where high interconnect density and low latency are required.

Micro-optical Packaging

Micro-optical packaging integrates optical components such as lenses, mirrors, and waveguides within the PIC package. Optical interconnects require precise alignment of optical fibers and waveguides to ensure efficient light transmission with minimal loss. Micro-lenses can be integrated to focus or collimate light beams, enhancing coupling efficiency and signal integrity. This packaging strategy is essential for photonic applications that demand high precision and low optical losses, such as data communications and sensing.

Advanced Packaging Materials

The choice of materials for packaging 3D PICs significantly impacts their performance and reliability. Advanced materials provide enhanced thermal, mechanical, and electrical properties, improving device efficiency.

High Thermal Conductivity Materials

Materials with high thermal conductivity, such as diamond and diamond-like carbon (DLC), are ideal for heat spreaders and substrates in high-power applications. Graphene-based thermal interface materials (TIMs) enhance heat transfer between components, improving overall thermal management. These materials are particularly beneficial for packaging high-power lasers, modulators, and other heat-intensive photonic components.

Low-Expansion Coefficient Materials

Materials with low thermal expansion coefficients help prevent mechanical stress and misalignment due to temperature variations. Kovar, an iron-nickel-cobalt alloy, has a thermal expansion coefficient similar to silicon, minimizing thermal mismatch. Alumina and aluminum nitride (AlN) ceramics provide low thermal expansion and high thermal conductivity, making them suitable for substrates and enclosures in environments with significant temperature fluctuations.

Optical Adhesives and Encapsulants

Optical adhesives and encapsulants protect and stabilize optical components within the PIC package. UV-curable adhesives cure rapidly under UV light, providing strong bonds with minimal shrinkage and high optical clarity. Epoxy-based encapsulants offer excellent mechanical protection and environmental resistance. These materials are essential for securing optical fibers, lenses, and other delicate components to ensure long-term reliability and performance.

Thermal Management in Packaging

Effective thermal management is crucial to prevent overheating and ensure the reliable operation of 3D PICs.

Heat Sinks and Spreaders

Heat sinks and spreaders dissipate heat away from the PIC, maintaining optimal operating temperatures. Integrated heat spreaders (IHS) distribute heat evenly across the device, while micro-heat sinks can be integrated directly into the PIC structure for enhanced local heat dissipation. These solutions are widely used in high-power applications such as data centers and telecommunications.

Microfluidic Cooling

Microfluidic cooling involves circulating coolant fluids through microchannels within the package to remove heat efficiently. Embedded microchannels etched into the substrate or package provide direct cooling to heat-generating components. Coolants such as water, ethylene glycol, and specialized fluids with high thermal conductivity are used for efficient heat transfer. This approach is ideal for applications with high heat flux, such as high-performance computing and advanced photonic devices.

Thermoelectric Coolers (TECs)

TECs use the Peltier effect to transfer heat from the PIC to a heat sink, providing active cooling. These modules consist of thermoelectric materials that generate a cooling effect when an electric current is applied. TECs offer precise temperature control and can be integrated into the package for localized cooling, making them valuable in applications where precise thermal management is critical, such as quantum photonic devices and sensitive optical sensors.

Case Studies and Applications

High-Power Lasers

High-power lasers generate significant heat, requiring robust packaging solutions to ensure performance and reliability. Diamond heat spreaders enhance thermal dissipation, preventing thermal degradation and ensuring stable operation. Hermetic packaging protects lasers from environmental contaminants, extending their lifespan and reliability.

Data Centers and Optical Interconnects

Data centers require efficient thermal management and high-performance packaging for optical interconnects. Flip-chip bonding provides efficient thermal and electrical connections, enhancing the performance of optical interconnects. Integrating microfluidic cooling channels within the package ensures effective heat dissipation, maintaining optimal operating temperatures.

Quantum Photonics

Quantum photonic devices are highly sensitive to environmental conditions, necessitating precise and robust packaging solutions. Low-expansion coefficient materials such as Kovar and AlN minimize thermal stress and maintain alignment precision. TECs provide precise temperature control, ensuring the stability and performance of quantum photonic circuits.

Biosensors and Medical Diagnostics

Biosensors and medical diagnostic devices require packaging solutions that ensure reliability and sensitivity. Optical adhesives, such as UV-curable adhesives, secure optical components with high precision, ensuring accurate measurements and long-term stability. Microfluidic cooling channels enhance thermal management, maintaining consistent operating conditions for sensitive biosensors

Future Directions

The future of packaging technologies for 3D PICs involves continuous innovation to address emerging challenges and enhance performance. The development of nanomaterials with tailored thermal, mechanical, and optical properties will provide new solutions for packaging 3D PICs. AI-driven design optimization using artificial intelligence and machine learning will enhance performance, reliability, and cost-effectiveness. Integrating multiple cooling techniques within a single package will improve thermal management and overall device performance.

Conclusion

Packaging technologies play a crucial role in the performance, reliability, and efficiency of 3D Photonic Integrated Circuits. Strategies such as hermetic packaging, flip-chip bonding, wire bonding, and through-silicon vias provide robust and efficient solutions for various applications. Advanced materials, including high thermal conductivity substances and low-expansion coefficient materials, enhance thermal management and structural integrity. Innovative cooling techniques, such as microfluidic cooling and thermoelectric coolers, ensure optimal operating temperatures and prevent thermal degradation. By leveraging these advanced packaging strategies, the challenges of 3D photonics can be overcome, ensuring the continued evolution and success of next-generation photonic devices.

6.3 Inter-layer Optical Interconnects in 3D Photonics: Solutions for Vertical Optical Connections with Low Loss and High Misalignment Tolerance

The evolution of 3D Photonic Integrated Circuits (PICs) has driven remarkable improvements in integration density, functionality, and performance by enabling the stacking of photonic layers in compact architectures. At the core of these advancements is the development of efficient inter-layer optical interconnects, which are essential for transferring optical signals vertically between stacked layers with minimal signal loss. Unlike electrical interconnects, optical interconnects must maintain high alignment tolerance and low propagation loss to ensure reliable communication between layers, especially as integration density increases. Various approaches, such as through-silicon vias (TSVs) and vertical grating couplers, have been explored to facilitate these connections. Vertical couplers are designed to direct light precisely between layers, minimizing loss while accommodating slight misalignments. Meanwhile, wafer bonding techniques—such as direct bonding and adhesive bonding—enable precise layer stacking with pre-aligned interconnects, enhancing overall robustness. Advanced materials, like silicon nitride and indium phosphide, are also employed to optimize the optical properties of inter-layer interconnects, ensuring efficient transmission at different wavelengths. This chapter delves into these technologies and explores the unique challenges and solutions associated with creating robust, efficient inter-layer optical interconnects for 3D photonic devices, which are crucial for high-performance applications in fields like data communication, quantum computing, and high-density photonic processing (Fig. 6.3).

Importance of Inter-layer Optical Interconnects

Inter-layer optical interconnects play a critical role in enabling the vertical integration of photonic components in 3D Photonic Integrated Circuits (PICs). They enhance integration density by allowing more components to be packed within a smaller footprint, increasing the overall scalability of photonic circuits. Efficient vertical interconnects also reduce

Fig. 6.3 Diagram of inter-layer optical interconnects, highlighting solutions for low-loss, high-misalignment-tolerance vertical connections. https://www.researchgate.net/figure/Cross-section-of-optical-interconnect-structure_fig1_3338003

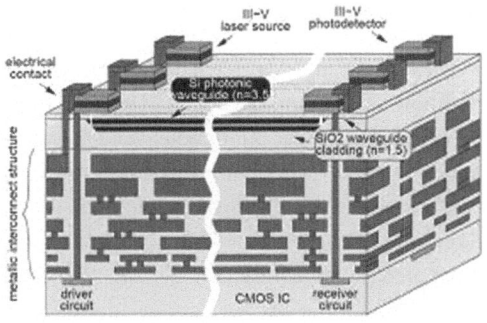

signal loss and latency, improving the overall performance of the PIC. Furthermore, they enable scalable designs, allowing for additional layers and functionalities without significantly increasing the chip size, making them essential for future photonic systems.

Challenges in Inter-layer Optical Interconnects

Developing effective inter-layer optical interconnects presents several challenges. One of the primary concerns is minimizing optical loss during signal transmission between layers to maintain signal integrity. High misalignment tolerance is also crucial, as slight fabrication errors or operational shifts can impact performance. Additionally, thermal management is a key challenge, as densely packed layers generate significant heat that must be efficiently dissipated to maintain stability and reliability.

Solutions for Inter-layer Optical Interconnects

Vertical Grating Coupler

Vertical grating couplers are widely used for coupling light between different layers in 3D PICs. These couplers use periodic structures to diffract light from one layer to another, facilitating efficient vertical interconnections. The grating period, depth, and duty cycle must be carefully designed to optimize coupling efficiency and minimize loss. Vertical grating couplers can be seamlessly integrated into existing fabrication processes, making them an attractive solution for multi-layer silicon photonics applications.

Micro-mirrors

Micro-mirrors provide a means of redirecting light vertically between layers by using reflective surfaces that can be tilted or curved to direct optical signals. These mirrors can be integrated into the PIC during fabrication or incorporated into the packaging process. Micro-mirrors offer precise control over light direction, enabling high coupling efficiencies. They are commonly used in optical switches, modulators, and other photonic devices where light needs to be routed between different layers or components.

Vertical Waveguides

Vertical waveguides offer a direct optical path between layers, ensuring low loss and high tolerance to misalignment. These waveguides are created using etching or deposition processes that form continuous optical pathways between layers. High-refractive-index materials, such as silicon or silicon nitride, are commonly used to ensure efficient light confinement and transmission. Vertical waveguides are ideal for applications requiring direct optical connections between layers, such as multi-layer optical interconnects and integrated photonic circuits.

Optical Vias

Optical vias are specialized structures that enable vertical optical connections between layers in 3D PICs. Designed to guide light vertically with minimal loss, these vias can be filled with low-loss optical materials or air. They are fabricated using precision etching and deposition techniques to ensure accurate alignment. Optical vias provide a direct and efficient means of vertical light transmission, reducing the complexity of optical signal routing. They are particularly useful in high-density vertical interconnects for data centers and advanced computing.

Plasmonic Interconnects

Plasmonic interconnects use surface plasmon polaritons (SPPs) to achieve vertical optical connections with high bandwidth and low loss. SPPs are electromagnetic waves that propagate along the interface between a metal and a dielectric, allowing for efficient optical transmission at nanoscale dimensions. The choice of materials and the geometry of plasmonic structures are critical to optimizing performance. These interconnects offer high bandwidth and low loss, making them suitable for high-speed optical communications and next-generation data centers.

Advanced Techniques for Misalignment Tolerance

Self-aligned Structures

Self-aligned structures naturally align during the fabrication process, reducing the risk of misalignment. Capillary forces during assembly can be leveraged to align components precisely, using solder or adhesive droplets that pull components into position as they solidify. Additionally, mechanical guides, such as V-grooves or alignment pins, ensure accurate positioning during assembly. These techniques are widely used in fiber-optic alignment and waveguide integration, where precision is crucial.

Active Alignment

Active alignment techniques use real-time feedback and adjustment during assembly to achieve precise positioning. Optical feedback allows monitoring of alignment accuracy, with adjustments made to maximize coupling efficiency and minimize loss. Precision robotic systems equipped with vision and feedback mechanisms can dynamically adjust component positioning to ensure optimal alignment. Active alignment is commonly used in the assembly of high-precision photonic devices, such as laser diodes, modulators, and detectors.

Adaptive Optics

Adaptive optics employ dynamic elements that adjust their properties to compensate for misalignment and other distortions. Deformable mirrors can change shape in response to control signals, allowing for real-time correction of optical paths. Similarly, tunable lenses can adjust their focal length or position to maintain alignment and focus. These techniques are particularly valuable in imaging systems and adaptive photonic circuits that require continuous alignment adjustments.

Thermal Management Considerations

Maintaining alignment and performance in inter-layer optical interconnects requires effective thermal management.

Heat Spreaders

Heat spreaders distribute heat evenly across the package to prevent hotspots. High thermal conductivity materials, such as diamond and copper, are commonly used to enhance heat dissipation. Heat spreaders are strategically designed to cover critical areas, ensuring uniform temperature distribution. They are widely used in high-power photonic devices where efficient heat dissipation is essential for maintaining performance and reliability.

Microfluidic Cooling

Microfluidic cooling systems circulate coolant fluids through microchannels to remove heat efficiently. Embedded microchannels are integrated into the PIC structure to provide localized cooling to heat-generating components. Common coolants include water, ethylene glycol, and specialized fluids with high thermal conductivity. This approach is ideal for applications with high heat flux, such as high-performance computing and advanced photonic devices.

Case Studies and Applications

Data Centers and Optical Interconnects

Data centers rely on efficient and high-performance optical interconnects to handle increasing data demands. Vertical grating couplers are used to achieve efficient coupling between stacked layers, reducing signal loss and latency. Plasmonic interconnects enable high-speed data transmission with low loss, enhancing the overall performance of optical networks.

Quantum Photonics

Quantum photonic devices require precise alignment and low-loss optical interconnects to maintain quantum coherence and signal integrity. Adaptive optics dynamically adjust optical paths in quantum photonic circuits, ensuring precise alignment and optimal performance. Optical vias provide low-loss vertical connections, preserving signal integrity in quantum photonic applications.

Biosensors and Medical Diagnostics

Biosensors and medical diagnostic devices require robust and reliable optical interconnects for accurate measurements and analysis. Micro-mirrors direct light between layers, enhancing coupling efficiency and reducing signal loss. Vertical waveguides provide efficient optical paths, ensuring accurate and reliable signal transmission in diagnostic devices.

Future Directions

The future of inter-layer optical interconnects in 3D photonics will involve continuous innovation to enhance performance and reliability. Developing nanophotonic structures with tailored optical properties will improve coupling efficiency and misalignment tolerance. Artificial intelligence and machine learning will play a crucial role in optimizing the design and placement of optical interconnects, enhancing performance while reducing development time. Additionally, integrating multiple cooling techniques within a single package will ensure efficient thermal management and maintain alignment precision, further advancing 3D photonic technologies.

Conclusion

Inter-layer optical interconnects are crucial for achieving vertical integration in 3D Photonic Integrated Circuits (PICs) by enabling the seamless transmission of optical signals between stacked photonic layers. This high-density integration requires interconnects that can efficiently transmit light while tolerating slight misalignments, as even minor deviations can significantly affect signal integrity and device performance. Vertical grating couplers use finely tuned gratings to direct light between layers with minimal loss, while micro-mirrors and vertical waveguides provide additional pathways to maintain controlled light propagation with reduced diffraction and scattering. Optical vias and plasmonic interconnects further optimize space by leveraging sub-wavelength dimensions, allowing light to be confined in highly compact structures, which is ideal for dense integration.

To maintain precise alignment, self-aligned structures automatically adjust components to correct minor shifts, whereas active alignment and adaptive optics use real-time feedback to dynamically correct positioning, ensuring optimal signal quality. Since thermal fluctuations can disrupt alignment and cause performance degradation, thermal management solutions such as heat spreaders and microfluidic cooling systems are integrated to keep temperatures stable, thereby preserving alignment precision and enhancing reliability. Together, these innovations in inter-layer interconnects and thermal management allow 3D PICs to achieve unprecedented levels of performance, scalability, and resilience for applications in telecommunications, high-speed data processing, and advanced sensing systems, driving the future of miniaturized, high-efficiency photonic technology.

Heterogeneous and Hybrid Integration

7

- **Combining Photonic and Electronic Materials**: Benefits and challenges of hybrid integration.
- **Applications in High-Tech Fields**: Impact on telecommunications, data centers, and advanced computing.
- **Hybrid Integration Techniques**: Methods to combine different material systems in a single device.

7.1 Combining Photonic and Electronic Materials: Benefits and Challenges of Hybrid Integration

The integration of photonic and electronic materials into a single platform, known as hybrid integration, represents a pivotal advancement for next-generation technologies. This approach harnesses the high-speed, low-loss data transmission capabilities of photonic components and combines them with the versatile processing power and control mechanisms of electronics, enabling significant improvements in performance, efficiency, and scalability across applications like data centers, telecommunications, and sensing systems. Hybrid integration allows for faster data rates, reduced latency, and enhanced energy efficiency, essential for addressing the demands of modern applications. However, achieving this seamless integration involves overcoming several challenges, including thermal management, material compatibility, and precise alignment of photonic and electronic components. Thermal issues arise as photonic components generate less heat than electronic ones, necessitating innovative cooling strategies to maintain stability. Additionally,

Fig. 7.1 Diagram showing the integration of photonic and electronic materials in a hybrid device. https://photonicsreport.com/blog/the-fascinating-relationship-between-photonics-and-electronics/

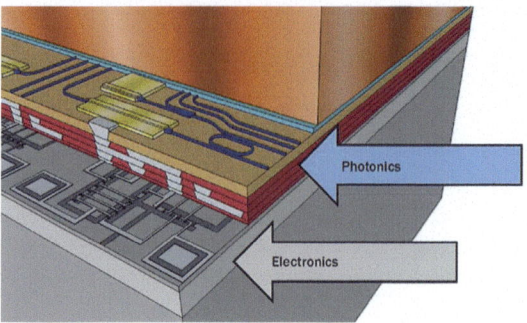

ensuring material compatibility—particularly between silicon-based electronics and various photonic materials like indium phosphide or silicon nitride—demands advanced fabrication techniques and bonding processes. This chapter provides an in-depth analysis of both the advantages and challenges of hybrid integration, illustrating how this transformative technology is setting the foundation for high-performance, compact, and multifunctional devices that will drive innovations in computing, healthcare, automotive technology, and beyond (Fig. 7.1).

Benefits of Hybrid Integration

Enhanced Performance and Functionality

The combination of photonic and electronic materials in hybrid integration enhances device performance by leveraging the strengths of both domains. Photonic components enable high-speed data transmission with low latency and minimal signal loss, making them ideal for communication networks and data centers. Electronic components provide advanced processing, control, and computational capabilities essential for complex signal processing and data management. The integration of these technologies allows for the development of compact systems that combine data transmission, processing, and control functions, significantly improving overall system performance.

Miniaturization and Integration Density

Hybrid integration facilitates the miniaturization of devices and increases integration density, leading to more compact and efficient systems. By integrating photonic and electronic components on a single chip, the overall footprint of the device is reduced, saving space and enabling more complex functionalities within the same area. This approach allows for higher integration density, enabling more components to be packed into a smaller area without compromising performance, making hybrid integration a key enabler for next-generation electronic-photonic systems.

Energy Efficiency

Photonic components are inherently more energy-efficient for data transmission compared to electronic components, leading to overall energy savings in hybrid systems. Photonics require less power for high-speed data transmission, reducing the overall power consumption of the device. Additionally, hybrid integration improves thermal management by distributing heat-generating components more effectively and utilizing efficient photonic data transmission to minimize heat generation.

Versatility and Flexibility

Hybrid integration offers significant versatility and flexibility in design, enabling the development of tailored solutions for specific applications. The ability to integrate various photonic and electronic components allows for customizable designs optimized for specific performance requirements. Hybrid systems can be applied across a wide range of fields, including telecommunications, data centers, sensing, medical diagnostics, and quantum computing.

Challenges of Hybrid Integration

Material Compatibility

Ensuring material compatibility between photonic and electronic components is a major challenge in hybrid integration. Different materials have varying thermal expansion coefficients, which can lead to mechanical stress and alignment issues due to temperature fluctuations. The quality of the interface between photonic and electronic materials is critical for efficient signal transmission and overall device performance. Imperfections or mismatches at the interface can result in signal loss and degraded performance.

Fabrication Complexity

The fabrication of hybrid integrated devices involves complex processes that require high precision and control. Achieving precise alignment between photonic and electronic components is crucial for optimal performance, as misalignment can lead to signal loss, crosstalk, and reduced efficiency. Developing reliable integration techniques that ensure strong bonding and efficient signal transmission between different materials is essential. This includes advanced lithography, etching, and deposition processes.

Thermal Management

Effective thermal management is critical in hybrid integrated systems to prevent overheating and ensure reliable operation. High-power electronic components generate significant heat, which must be dissipated effectively to prevent thermal degradation of both electronic and photonic components. Maintaining consistent temperatures across the hybrid device is necessary to prevent thermal-induced stresses and alignment issues that could affect performance and reliability.

Signal Integrity

Maintaining signal integrity in hybrid integrated systems is essential for ensuring high performance and reliability. Minimizing optical loss at the interfaces and throughout the photonic components is crucial for maintaining strong signal integrity. Additionally, managing electrical noise and crosstalk between electronic components is critical to ensure clean and reliable signal processing.

Key Technologies for Hybrid Integration

Silicon Photonics

Silicon photonics is a key technology for hybrid integration, leveraging well-established silicon fabrication processes used in the electronics industry. Its compatibility with complementary metal-oxide-semiconductor (CMOS) fabrication processes enables seamless integration with electronic components. Silicon photonics allows for high integration density of photonic components, such as waveguides, modulators, and detectors, on a single chip.

III–V Semiconductors

III–V semiconductors, such as indium phosphide (InP) and gallium arsenide (GaAs), offer superior optoelectronic properties compared to silicon. These materials are used to fabricate high-efficiency lasers and photodetectors, which are essential for high-performance optical communication systems. Advanced bonding and epitaxial growth techniques enable the integration of III–V materials with silicon, combining the benefits of both material systems to improve hybrid integration capabilities.

Heterogeneous Integration Techniques

Heterogeneous integration involves combining different material systems and components onto a single substrate or package. Direct wafer bonding techniques enable the integration of different materials by bonding them at the atomic level, ensuring strong and reliable interfaces. Selective area growth techniques involve growing III–V materials directly on silicon substrates in predefined areas, allowing for precise placement and integration of photonic components.

Advanced Packaging

Advanced packaging techniques are essential for protecting hybrid integrated devices and ensuring their performance and reliability. Flip-chip bonding provides efficient electrical and thermal connections between the chip and the substrate, enhancing performance and thermal management. Micro-optical packaging techniques integrate optical components, such as lenses and waveguides, within the package to ensure efficient light coupling and minimal loss.

Case Studies and Applications

Data Centers and High-Speed Communication

Hybrid integration is transforming data centers and high-speed communication networks by enabling efficient and high-performance optical interconnects. Silicon photonic transceivers integrate photonic components, such as modulators and detectors, with electronic control circuits, providing high-speed data transmission with low power consumption. Hybrid optical switches combine photonic switching elements with electronic control circuits, enabling fast and efficient data routing in data centers.

Medical Diagnostics and Biosensing

Hybrid integration is advancing medical diagnostics and biosensing by enabling compact, high-performance devices for detecting and analyzing biological signals. Integrated biosensors combine photonic sensing elements with electronic processing circuits, providing high sensitivity and real-time data analysis for medical diagnostics. Lab-on-a-chip devices integrate photonic and electronic components to perform complex biochemical analyses on a single chip, reducing the need for large and expensive laboratory equipment.

Quantum Computing and Communications

Hybrid integration plays a crucial role in the development of quantum computing and communication technologies. Quantum photonic circuits integrate single-photon sources, waveguides, and detectors with electronic control circuits, enabling the manipulation and measurement of quantum states. Hybrid integrated devices facilitate the high-speed and secure transmission of quantum information, essential for developing robust quantum communication networks.

Future Directions

The future of hybrid integration in photonics and electronics will involve continuous innovation to address existing challenges and enhance performance. The development of nanophotonic structures with tailored optical properties will improve integration density and performance. Artificial intelligence and machine learning will play a key role in optimizing the design and fabrication processes, reducing development time and improving efficiency. Research into advanced materials with improved thermal, mechanical, and optical properties will further enhance hybrid integration capabilities, leading to more reliable and efficient photonic-electronic systems.

Conclusion

Hybrid integration of photonic and electronic materials offers significant benefits, including enhanced performance, miniaturization, energy efficiency, and versatility. However, it also presents unique challenges, such as material compatibility, fabrication complexity,

Fig. 7.2 Application map showcasing the impact of hybrid integration in telecommunications, data centers, and advanced computing. https://www.snsin.com/what-is-a-hybrid-data-center/

thermal management, and signal integrity. Key technologies such as silicon photonics, III–V semiconductors, heterogeneous integration techniques, and advanced packaging play crucial roles in overcoming these challenges. By leveraging these technologies, hybrid integration can drive innovation and enable the development of next-generation devices for a wide range of applications, from data centers and high-speed communication to medical diagnostics and quantum computing (Fig. 7.2).

The integration of photonic and electronic materials into advanced hybrid systems has profound implications for several high-tech fields, including telecommunications, data centers, and advanced computing. This chapter explores how these sectors are leveraging the benefits of photonic integrated circuits (PICs) and hybrid integration to enhance performance, efficiency, and scalability.

Telecommunications

Telecommunications is one of the most significant beneficiaries of advancements in photonic integration. The demand for higher data rates, increased bandwidth, and lower latency continues to drive the adoption of photonic technologies in this sector.

High-Speed Optical Communication

Hybrid integration of photonic and electronic components enables the development of high-speed optical communication systems, which are essential for meeting the growing demand for data transmission. Photonic transceivers that integrate lasers, modulators, and photodetectors with electronic control circuits provide high-speed data transmission with low latency. These devices are critical for fiber-optic communication networks, enabling data rates of up to 400 Gbps and beyond. Dense Wavelength Division Multiplexing (DWDM) technology, which uses multiple wavelengths to transmit data over a single optical fiber, benefits from hybrid integrated PICs. This approach significantly increases the capacity of optical communication networks, supporting the ever-growing data traffic.

Signal Processing and Switching

Photonic integration plays a crucial role in enhancing signal processing and switching capabilities in telecommunications. Hybrid integrated optical switches, which combine photonic switching elements with electronic control circuits, offer fast and efficient data routing. These switches reduce signal latency and power consumption compared to electronic switches, enhancing network performance. Photonic signal regeneration, which involves the use of hybrid integrated amplifiers and wavelength converters, ensures signal integrity over long distances. This capability is essential for maintaining high-quality communication in long-haul networks.

5G and Beyond

The deployment of 5G networks and the development of future wireless technologies rely heavily on photonic integration. High-capacity optical links are essential for connecting 5G base stations to the core network (backhaul) and for distributing signals from the base stations to remote radio heads (fronthaul). Hybrid integrated PICs provide the high data rates and low latency required for these links. Additionally, photonic integration enables the creation of virtualized network slices, each tailored for specific applications and services. This capability supports the diverse requirements of 5G and future networks, from enhanced mobile broadband to ultra-reliable low-latency communications.

Data Centers

Data centers are at the heart of the digital economy, providing the infrastructure for cloud computing, big data analytics, and internet services. Photonic integration significantly enhances the performance, efficiency, and scalability of data centers.

Optical Interconnects

Optical interconnects are crucial for achieving high-speed, low-latency communication within data centers. Hybrid integrated PICs enable high-speed optical links between servers, switches, and storage devices within data centers. These interconnects reduce latency and increase bandwidth, supporting the high-performance requirements of modern data centers. Optical interconnects also facilitate efficient communication between different racks in a data center, minimizing signal degradation and power consumption compared to traditional electrical interconnects.

Energy Efficiency

Energy efficiency is a critical concern for data centers, which consume significant amounts of power. Hybrid integrated photonic devices consume less power than their electronic counterparts for data transmission, reducing the overall energy consumption of data centers. Efficient thermal management techniques, such as microfluidic cooling integrated with photonic components, help dissipate heat more effectively, further enhancing energy efficiency.

Scalability and Flexibility

Photonic integration provides the scalability and flexibility needed to meet the growing demands on data center infrastructure. Hybrid integrated PICs support modular data center architectures, allowing for easy expansion and upgrades. This modularity ensures that data centers can scale efficiently to accommodate increasing data traffic. Photonic switches and reconfigurable optical add-drop multiplexers (ROADMs) enable dynamic reconfiguration of data center networks. This flexibility allows data centers to adapt to changing workloads and optimize resource utilization.

Advanced Computing

Advanced computing applications, including high-performance computing (HPC), artificial intelligence (AI), and quantum computing, benefit significantly from photonic integration.

High-Performance Computing

HPC systems require ultra-fast data transfer rates and low-latency communication to handle complex computations. Hybrid integrated photonic interconnects provide the high bandwidth and low latency required for HPC systems. These interconnects enable efficient communication between processing units, memory modules, and storage devices, enhancing overall system performance. The integration of photonic components with electronic processors enables the development of photonic processors, which offer higher data processing speeds and lower power consumption than traditional electronic processors.

Artificial Intelligence

AI applications demand substantial computational power and efficient data handling. Hybrid integrated neuromorphic chips, which mimic the structure and function of the human brain, leverage photonic integration to achieve high-speed data transfer and parallel processing. These chips are essential for AI applications that require real-time data processing and learning. Optical neural networks, which use photonic components for data transmission and processing, offer significant advantages in terms of speed and energy efficiency. Hybrid integration enables the development of compact and efficient optical neural networks for AI applications.

Quantum Computing

Quantum computing represents a new paradigm in computing, with the potential to solve problems that are intractable for classical computers. Hybrid integrated quantum photonic circuits combine single-photon sources, waveguides, and detectors with electronic control circuits. These circuits are essential for manipulating and measuring quantum states, enabling the development of scalable quantum computers. Secure quantum communication relies on the efficient transmission and detection of quantum information.

Hybrid integrated devices provide the high-speed and secure transmission of quantum information, supporting the development of robust quantum communication networks.

Challenges and Future Directions

Integration Complexity

The integration of photonic and electronic materials into hybrid systems involves complex fabrication processes and precise alignment. Developing reliable fabrication techniques that ensure precise alignment and strong bonding between photonic and electronic components is essential. Advanced lithography, etching, and deposition processes play a critical role in achieving this precision. Ensuring the scalability of hybrid integration techniques to accommodate increasing complexity and integration density is crucial for future applications.

Material Compatibility

Ensuring compatibility between different materials used in hybrid integration is a significant challenge. Different materials have varying thermal expansion coefficients, which can lead to mechanical stress and alignment issues during temperature fluctuations. Developing materials with matched thermal properties is essential for reliable hybrid integration. High-quality interfaces between photonic and electronic materials are critical for efficient signal transmission and overall device performance. Research into advanced bonding and interface engineering techniques is ongoing to address this challenge.

Thermal Management

Effective thermal management is essential to prevent overheating and ensure the reliable operation of hybrid integrated devices. High-power electronic components generate significant heat, which must be dissipated effectively to prevent thermal degradation of both electronic and photonic components. Developing advanced cooling techniques, such as microfluidic cooling and thermoelectric coolers, integrated with photonic components, is crucial for maintaining optimal operating temperatures.

Future Directions

The future of hybrid integration in high-tech fields involves continuous innovation to enhance performance, scalability, and reliability. The development of nanophotonic structures with tailored optical properties will improve integration density and performance. Artificial intelligence and machine learning will play a key role in optimizing the design and fabrication processes, reducing development time and improving efficiency. Research into advanced materials with improved thermal, mechanical, and optical properties will further enhance hybrid integration capabilities, leading to more reliable and efficient photonic-electronic systems.

Conclusion

The integration of photonic and electronic materials into hybrid systems offers significant benefits for telecommunications, data centers, and advanced computing. High-speed optical communication, enhanced signal processing, and efficient data transmission are just a few of the advantages that photonic integration brings to these high-tech fields. However, challenges related to material compatibility, fabrication complexity, and thermal management must be addressed to fully realize the potential of hybrid integration. By leveraging advanced technologies and continuous innovation, hybrid integration can drive the development of next-generation devices and systems, transforming the landscape of telecommunications, data centers, and advanced computing.

7.2 Hybrid Integration Techniques: Methods to Combine Different Material Systems in a Single Device

The rapid advancement of photonic and electronic technologies has driven the need for hybrid integration techniques that combine multiple material systems within a single device, allowing for optimal use of each material's unique properties. These techniques enable the merging of materials such as silicon, which supports scalable electronics, with photonic materials like indium phosphide or silicon nitride, which excel in light manipulation. Hybrid integration leverages each material's strengths—such as the speed and data capacity of photonics and the processing capabilities of electronics—leading to devices with enhanced performance, functionality, and scalability. Achieving this integration, however, requires sophisticated methods, including wafer bonding, where entire wafers of different materials are bonded to create seamless layers, and flip-chip bonding, which facilitates high-density connections between photonic and electronic layers. Heterogeneous integration also plays a key role, where different material components are assembled on a shared platform, preserving their individual properties while enhancing overall device performance. This chapter provides a comprehensive look at these integration techniques, discussing how they allow for multi-functional, compact, and high-performance devices that are critical for applications in data communications, sensor technology, and next-generation computing systems (Fig. 7.3).

Overview of Hybrid Integration

Hybrid integration refers to the combination of different materials and components into a single, cohesive system. This approach allows designers to utilize the best properties of each material, resulting in devices with superior performance and capabilities.

Advantages of Hybrid Integration

Hybrid integration offers several key advantages. By combining materials with complementary properties, it enhances overall device performance and allows for the

7.2 Hybrid Integration Techniques: Methods to Combine Different …

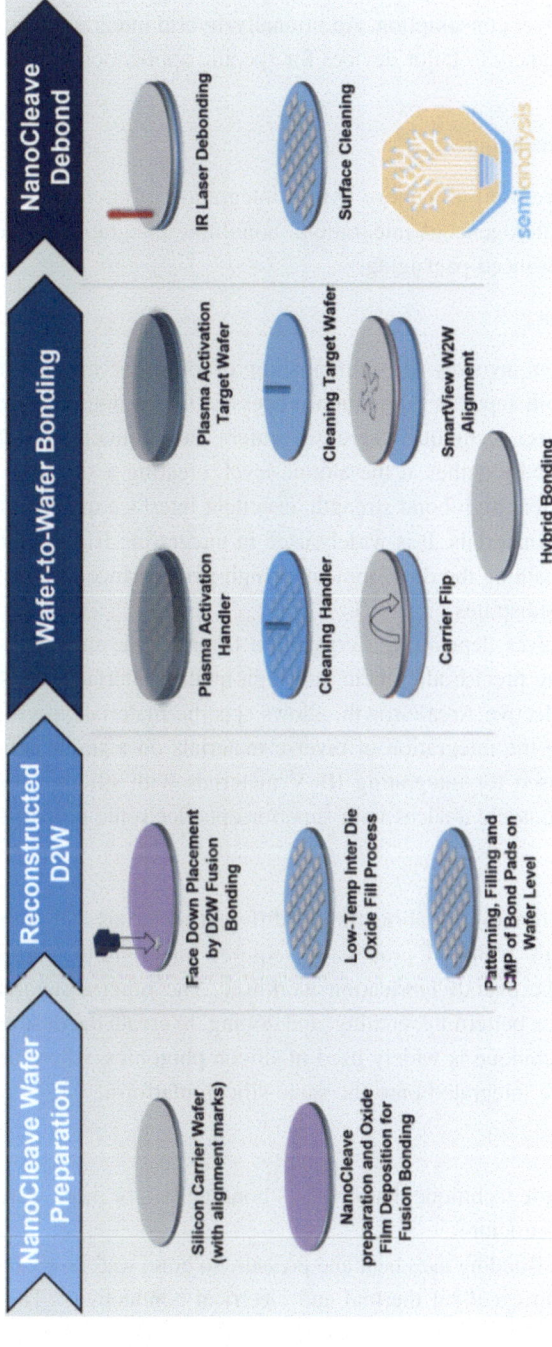

Fig. 7.3 Illustration of different hybrid integration techniques, such as wafer bonding and flip-chip bonding. https://www.semianalysis.com/p/hybrid-bonding-process-flow-advanced

incorporation of diverse functionalities within a single system. This approach supports scalability, enabling the development of more complex and integrated devices without increasing footprint or power consumption. Additionally, hybrid integration facilitates customization, allowing designers to tailor devices for specific applications and performance requirements.

Key Hybrid Integration Techniques

Several key techniques are used to achieve hybrid integration in photonic and electronic devices. These include heterogeneous integration, monolithic integration, wafer bonding, flip-chip bonding, and advanced packaging.

Heterogeneous Integration

Heterogeneous integration involves the combination of different material systems or components, typically from separate fabrication processes, into a single device.

Direct Wafer Bonding is a technique where two wafers, each containing different materials or devices, are bonded together at the atomic level, creating a strong and seamless interface. This method offers high bond strength, excellent interface quality, and the ability to integrate dissimilar materials. It is widely used in integrating III–V semiconductors with silicon photonics, enabling the development of high-performance lasers, modulators, and detectors on silicon substrates.

Epitaxial Growth involves depositing a crystalline layer of one material on the substrate of another, ensuring precise alignment and high-quality interfaces. A key method within this technique, Selective Area Growth, allows specific materials to grow in designated regions, facilitating the integration of diverse materials on a single substrate. This approach is commonly used for integrating III–V materials with silicon, supporting the development of hybrid photonic devices with superior optoelectronic properties.

Monolithic Integration

Monolithic integration refers to the fabrication of different materials and components on a single substrate through a unified process. It requires materials that are compatible with the substrate and the overall fabrication workflow. The process involves multiple steps, including deposition, patterning, etching, and doping, to create the desired structures and components. This technique is widely used in silicon photonics, where photonic and electronic components are integrated onto the same silicon platform.

Wafer Bonding

Wafer bonding is a versatile technique that involves bonding two or more wafers together to create a multi-layered structure.

Thermal Compression Bonding uses heat and pressure to bond wafers, creating a strong and durable interface with excellent thermal and electrical conductivity. This method is frequently applied in fabricating 3D integrated circuits and MEMS devices.

Adhesive Bonding utilizes an adhesive layer to bond wafers, making it suitable for materials that cannot withstand high temperatures. This approach allows for low-temperature processing and greater flexibility in material selection. It is commonly used in integrating photonic and electronic components, such as in optoelectronic packaging.

Flip-Chip Bonding

Flip-chip bonding is a technique where a chip is mounted upside-down on a substrate, allowing for direct electrical and thermal contact.

Bump Technology employs metal bumps, typically made of solder or gold, deposited on the chip's bonding pads to provide electrical connections and mechanical support. Underfill materials are used to enhance mechanical stability and thermal performance. This method reduces interconnect lengths, lowers electrical resistance, and improves thermal dissipation, making it widely used in high-performance computing, telecommunications, and optoelectronic devices.

Advanced Packaging

Advanced packaging techniques are crucial for protecting and enhancing the performance of hybrid integrated devices.

System-in-Package (SiP) integrates multiple chips and components into a single package, combining different material systems and functionalities. This method provides a compact size, reduced signal loss, and enhanced performance, making it ideal for mobile devices, IoT applications, and high-performance computing.

Fan-Out Wafer-Level Packaging (FOWLP) extends the chip's footprint by redistributing interconnects, enabling the integration of additional components and materials. This technique supports high integration density, improves thermal management, and reduces form factor, making it suitable for advanced mobile processors, RF modules, and sensors.

Challenges in Hybrid Integration

Despite its benefits, hybrid integration presents several challenges that must be addressed to ensure successful implementation.

Material Compatibility

Ensuring compatibility between different materials is critical for the performance and reliability of hybrid integrated devices. Thermal expansion mismatches between materials can lead to mechanical stress and alignment issues during temperature fluctuations. High-quality interfaces are essential for efficient signal transmission, and techniques such as direct wafer bonding and epitaxial growth are used to improve interface quality.

Fabrication Complexity

The fabrication of hybrid integrated devices involves complex processes that require precise control and alignment. Achieving precision alignment is crucial, as even minor misalignments can lead to signal loss and reduced efficiency. Advanced lithography, etching, and deposition techniques are essential for ensuring alignment accuracy. Additionally, integrating multiple fabrication processes for different materials poses a significant challenge, requiring the development of scalable and reliable integration techniques.

Thermal Management

Effective thermal management is critical to prevent overheating and ensure reliable operation. Heat dissipation is a significant concern, particularly for high-power components, which generate substantial heat. Advanced cooling techniques, such as microfluidic cooling and thermoelectric coolers integrated with photonic components, are essential for maintaining optimal operating temperatures.

Applications of Hybrid Integration

Telecommunications

Hybrid integration is transforming telecommunications by enabling high-speed optical communication systems and advanced signal processing capabilities. High-speed transceivers that integrate photonic components with electronic control circuits provide fast, low-latency data transmission, essential for modern communication networks. Optical switches leveraging hybrid integration offer efficient data routing with minimal power consumption.

Data Centers

Data centers benefit significantly from hybrid integration through enhanced performance, energy efficiency, and scalability. Optical interconnects support high-speed, low-latency communication within data centers, improving infrastructure efficiency. Hybrid integrated photonic devices also contribute to energy efficiency, reducing power consumption compared to electronic-only systems.

Advanced Computing

Advanced computing applications, including high-performance computing (HPC) and artificial intelligence (AI), leverage hybrid integration for enhanced computational power and efficiency. Photonic processors, which integrate photonic components with electronic processors, offer faster data processing with lower power consumption. Optical neural networks provide significant advantages in speed and energy efficiency, supporting AI-driven applications.

Future Directions

- The future of hybrid integration in photonics and electronics is marked by continuous innovation aimed at overcoming current challenges and pushing the boundaries of device performance, functionality, and scalability. Nanophotonic structures are at the forefront, with the development of ultra-compact, highly integrated designs tailored to manipulate light at the nanoscale. These structures, such as photonic crystals, plasmonic waveguides, and metasurfaces, enable higher integration densities and enhanced optical properties, allowing for more compact and efficient devices with minimized energy loss and improved bandwidth. By precisely controlling light at small scales, nanophotonic structures hold the potential to significantly improve data transfer rates and energy efficiency in hybrid photonic-electronic systems.

- AI-driven design is another promising avenue, leveraging artificial intelligence (AI) and machine learning (ML) to optimize design and fabrication processes. These technologies can analyze vast datasets to identify optimal configurations and predict performance outcomes, thus accelerating innovation. AI can streamline complex design processes, allowing for the creation of customized, high-performance devices that are not only more efficient but also quicker to develop. Furthermore, AI-driven tools can adapt designs based on real-time feedback during the fabrication process, reducing material waste and enhancing manufacturing precision.
- Advanced materials play a pivotal role in enhancing hybrid integration by addressing limitations in thermal management, mechanical stability, and optical efficiency. Research is underway to discover and develop materials that combine high thermal conductivity with durability and specific optical properties. For instance, materials like silicon carbide, diamond, and certain chalcogenides offer superior thermal properties, which are crucial in densely packed hybrid devices where heat dissipation is a challenge. Additionally, materials with adjustable refractive indices or high nonlinear optical coefficients can enable greater control over light manipulation and signal processing. These advances in material science will be crucial for creating hybrid integrated systems that are scalable, reliable, and versatile for applications in telecommunications, sensing, and quantum computing. As these innovations converge, the future of hybrid integration will see faster, more compact, and energy-efficient devices, paving the way for a new era in photonic and electronic integration.

Hybrid integration techniques are pivotal in combining diverse material systems into a single device, enabling substantial improvements in performance, functionality, and scalability. Techniques like heterogeneous integration allow different materials (such as silicon, indium phosphide, and silicon nitride) to be combined on a shared substrate, optimizing each material's unique properties for specific functions within the device. Monolithic integration, in contrast, incorporates different functionalities within a single material platform, which simplifies fabrication but often requires innovative design to balance performance across photonic and electronic components. Wafer bonding and flip-chip bonding are essential for connecting layers with precise alignment and low loss, supporting high-density integration and compact device architectures. Advanced packaging strategies ensure that these hybrid systems are mechanically robust and thermally managed, which is especially critical in applications like telecommunications, data centers, and advanced computing where high performance and reliability are paramount.

While material compatibility, fabrication complexity, and thermal management remain significant challenges, ongoing advancements in bonding techniques, cooling solutions, and material science are continuously expanding the potential of hybrid integration. Leveraging these diverse hybrid integration methods allows for the development of next-generation devices that meet the increasing demands of high-tech industries, paving the way for breakthroughs in high-speed data transmission, efficient energy use, and miniaturized yet powerful computational systems. This convergence of technologies promises to address the scalability and efficiency needs of modern applications, enabling powerful, compact, and multifunctional devices for future innovations.

Applications of 3D Photonics

8

- **Telecommunications and Data Centers**: Enhancing data transmission and processing capabilities.
- **Solid-State LiDAR and AI**: Improving performance and miniaturization for sensing and artificial intelligence applications.
- **Quantum Computing and Sensing**: Role of 3D PICs in advancing quantum technologies.

8.1 Telecommunications and Data Centers: Enhancing Data Transmission and Processing Capabilities Utilizing 3D Photonics

The integration of 3D photonics into telecommunications and data centers marks a transformative advancement, enabling far greater data transmission speeds, processing capabilities, and energy efficiency compared to traditional 2D photonic solutions. By utilizing vertical stacking of photonic components, 3D photonics allows for denser integration and shorter interconnects, reducing latency and minimizing signal loss—a crucial improvement for the high-speed demands of data centers and telecommunication networks.

The integration of 3D photonics into telecommunications and data centers marks a transformative advancement, enabling far greater data transmission speeds, processing capabilities, and energy efficiency compared to traditional 2D photonic solutions. By utilizing vertical stacking of photonic components, 3D photonics allows for denser integration and shorter interconnects, reducing latency and minimizing signal loss—a

© The Author(s), under exclusive license to Springer Nature Switzerland AG 2026
Y. Yi, *From 2D to 3D Photonic Integrated Circuits*, Synthesis Lectures on Emerging Engineering Technologies, https://doi.org/10.1007/978-3-031-91508-6_8

crucial improvement for the high-speed demands of data centers and telecommunication networks. 3D photonics technology leverages innovations like integrated waveguides, multiplexers, modulators, and photodetectors to achieve seamless optical signal transmission across layers, improving data rates and enabling more efficient data processing. Additionally, 3D photonic integration supports parallel processing and allows for scalable architecture, making it possible to handle larger data loads in a smaller footprint, which is essential as data centers continue to expand to meet global digital demands. By reducing power consumption and improving thermal management through compact designs, 3D photonics significantly reduces operational costs and environmental impact, offering a sustainable solution for next-generation telecommunications and data centers. This chapter examines these technological advancements, outlining the benefits, challenges, and real-world applications that are driving this leap forward in data communication and processing infrastructure (Fig. 8.1).

Overview of 3D Photonics

3D photonics utilizes three-dimensional structures and integration techniques to build advanced photonic devices and systems that extend beyond the limitations of traditional two-dimensional designs. By leveraging the third dimension, 3D photonics enables increased integration density, allowing for a higher concentration of photonic components in a compact space, which is essential for applications requiring high data throughput and minimal footprint, such as data centers and telecommunications. This vertical stacking of photonic elements—such as waveguides, modulators, and detectors—reduces the distance that optical signals need to travel, which improves performance by minimizing

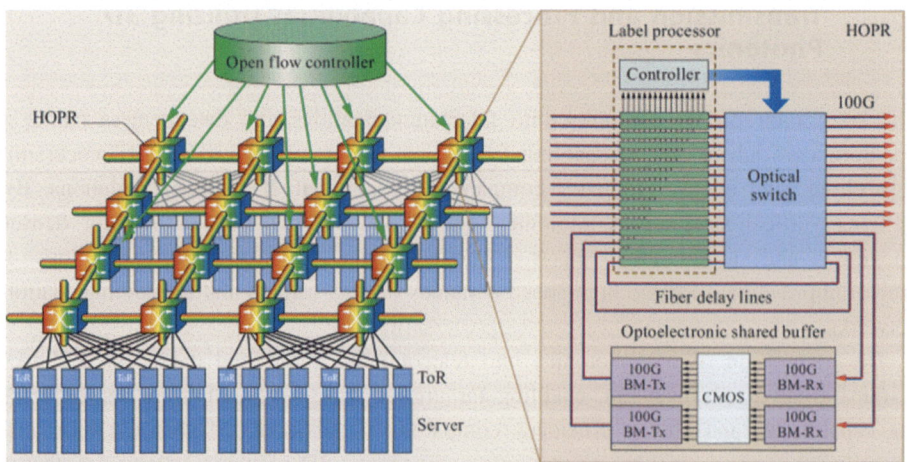

Fig. 8.1 Diagram of a 3D photonic switch for telecommunications, showing pathways for high-speed data transmission. https://link.springer.com/chapter/10.1007/978-3-030-16250-4_25

signal loss, enhancing speed, and reducing latency. Additionally, 3D photonic structures provide enhanced functionality through more complex architectures, enabling parallel processing and multi-layered signal routing that would be challenging to achieve in a 2D layout. These advancements facilitate the development of multifunctional, high-speed, and energy-efficient devices capable of meeting the demands of next-generation technology, from high-performance computing to sensing and communication systems. Overall, 3D photonics is paving the way for a new era of miniaturized, high-performance photonic systems that are both scalable and adaptable to future innovations in the field.

Key Advantages of 3D Photonics

3D photonics enables the vertical stacking of photonic components, significantly increasing the number of elements that can be integrated into a single device. This approach enhances performance by allowing more efficient light routing and signal processing, reducing losses, and improving overall device functionality. The scalability of 3D photonics makes it ideal for complex systems, such as those found in telecommunications and data centers, where multiple functionalities need to be seamlessly integrated.

Enhancing Data Transmission in Telecommunications

Telecommunications networks form the backbone of modern communication systems, and 3D photonics is playing a crucial role in improving their efficiency and capacity.

High-Speed Optical Communication

High-speed optical communication systems benefit significantly from 3D photonic integration, which enables devices with higher bandwidth and lower latency. 3D photonic transceivers integrate multiple photonic components, such as lasers, modulators, and detectors, into a single 3D structure. This integration enhances data rates and signal integrity, supporting speeds of up to 400 Gbps and beyond, which are essential for modern telecommunications networks. Dense Wavelength Division Multiplexing (DWDM), which uses multiple wavelengths to transmit data over a single optical fiber, also benefits from 3D photonic integration. The increased integration density allows for more channels within a single device, significantly boosting the capacity of optical networks.

Signal Processing and Switching

3D photonics enhances signal processing and switching capabilities, which are critical for efficient data routing and network management. 3D optical switches combine photonic switching elements with electronic control circuits, enabling fast and efficient data routing while reducing signal latency and power consumption. Advanced photonic signal processing techniques, such as optical signal regeneration, wavelength conversion, and coherent detection, ensure signal integrity over long distances, maintaining high-quality communication in long-haul networks.

5G and Future Wireless Technologies

The deployment of 5G networks and the development of future wireless technologies rely on 3D photonics to meet the growing demand for data transmission and connectivity. High-capacity optical links are essential for connecting 5G base stations to the core network (backhaul) and for distributing signals from base stations to remote radio heads (fronthaul). 3D photonic integration supports these links with high data rates and low latency, ensuring seamless connectivity. Additionally, 3D photonics enables the creation of virtualized network slices, each optimized for specific applications and services. This capability supports diverse 5G and future network requirements, ranging from enhanced mobile broadband to ultra-reliable low-latency communications.

Enhancing Data Processing in Data Centers

Data centers play a critical role in cloud computing, big data analytics, and internet services. 3D photonics significantly enhances their performance, efficiency, and scalability.

Optical Interconnects

Optical interconnects are essential for achieving high-speed, low-latency communication within data centers. 3D photonic integration enables high-speed optical links between servers, switches, and storage devices, reducing latency and increasing bandwidth. Rack-to-rack communication is also optimized through 3D photonics, facilitating efficient data transmission while minimizing signal degradation and power consumption compared to traditional electrical interconnects.

Energy Efficiency

Energy efficiency is a major concern for data centers, which consume significant amounts of power. 3D photonics contributes to lower power consumption, as photonic devices require less energy than their electronic counterparts for data transmission. Efficient thermal management techniques, such as microfluidic cooling integrated with 3D photonic components, further improve energy efficiency by effectively dissipating heat.

Scalability and Flexibility

The scalability and flexibility of 3D photonics allow data centers to adapt to increasing workloads and network demands. Modular architectures enabled by 3D photonic integration facilitate easy expansion and upgrades, ensuring that data centers can scale efficiently to accommodate growing data traffic. Reconfigurable networks, supported by photonic switches and reconfigurable optical add-drop multiplexers (ROADMs), allow for dynamic reconfiguration, optimizing resource utilization and adapting to changing workloads.

Advanced Computing Applications

3D photonic integration is transforming advanced computing applications, including high-performance computing (HPC), artificial intelligence (AI), and quantum computing.

High-Performance Computing

HPC systems require ultra-fast data transfer rates and low-latency communication to handle complex computations. 3D photonic interconnects provide the necessary high bandwidth and low latency, enabling efficient communication between processing units, memory modules, and storage devices. The integration of photonic components with electronic processors allows for the development of photonic processors, which offer faster data processing speeds and lower power consumption than traditional electronic processors.

Artificial Intelligence

AI applications demand substantial computational power and efficient data handling, and 3D photonics offers significant advantages in this area. Neuromorphic computing, which mimics the structure and function of the human brain, benefits from 3D photonic integration by achieving high-speed data transfer and parallel processing. Optical neural networks, which use photonic components for data transmission and processing, provide superior speed and energy efficiency. 3D photonic integration enables the development of compact and highly efficient optical neural networks, supporting AI applications that require real-time data processing.

Quantum Computing

Quantum computing represents a new paradigm in computational technology, with the potential to solve problems that are intractable for classical computers. 3D photonic integration facilitates the development of quantum photonic circuits by integrating single-photon sources, waveguides, and detectors with electronic control circuits. These circuits are essential for manipulating and measuring quantum states, enabling the development of scalable quantum computers. Secure quantum communication also relies on efficient quantum information transmission, which is enhanced by 3D photonic integration. These advancements support the creation of robust quantum communication networks for secure data exchange.

Future Directions

The future of 3D photonics in telecommunications, data centers, and advanced computing involves continuous innovation to enhance performance, scalability, and reliability. Researchers are developing nanophotonic structures with tailored optical properties to improve integration density and device efficiency. AI-driven design is being leveraged to optimize the fabrication and performance of 3D photonic devices, reducing development time and enhancing precision. Additionally, advanced materials with improved thermal, mechanical, and optical properties are being researched to further enhance the capabilities of 3D photonic integration.

Conclusion

3D photonics is revolutionizing telecommunications, data centers, and advanced computing by enhancing data transmission and processing capabilities. High-speed optical communication, advanced signal processing, and efficient data handling are just a few of the advantages that 3D photonic integration brings to these high-tech fields. While challenges related to material compatibility, fabrication complexity, and thermal management remain, continuous innovation and technological advancements are paving the way for the successful integration of 3D photonics. By leveraging these developments, the industry can create next-generation devices and systems that meet the growing demands of modern telecommunications, data centers, and computing applications.

8.2 Solid-State LiDAR and AI Using 3D Photonics: Improving Performance and Miniaturization for Sensing and Artificial Intelligence Applications

The emergence of 3D photonics has brought transformative advancements to high-tech fields such as solid-state LiDAR (Light Detection and Ranging) and artificial intelligence (AI), enabling new levels of performance, precision, and miniaturization that were previously unattainable. In solid-state LiDAR, 3D photonic integration enhances the ability to create compact, robust, and highly sensitive devices for depth sensing and environmental mapping. By using 3D photonics, LiDAR systems can achieve higher resolution, faster response times, and improved range without relying on mechanical parts, making them ideal for autonomous vehicles, robotics, and high-precision imaging applications. In AI, 3D photonic integration supports the development of high-speed optical neural networks and other AI hardware by enabling dense interconnections and low-latency data processing, which are critical for real-time, data-intensive computations. The stacked architecture of 3D photonics allows for parallel processing and closer integration of photonic and electronic components, significantly improving data throughput and computational efficiency. This chapter delves into the technological innovations that 3D photonics brings to LiDAR and AI, including advanced materials, waveguide design, and multi-layer photonic circuits. By examining the applications and benefits of 3D photonics, this chapter highlights how this technology is driving breakthroughs in sophisticated sensing, AI processing, and compact, scalable device design that are reshaping fields such as autonomous navigation, machine learning, and real-time environmental analysis (Fig. 8.2).

Fig. 8.2 Illustration of a solid-state LiDAR system with 3D photonic components for autonomous driving and AI applications

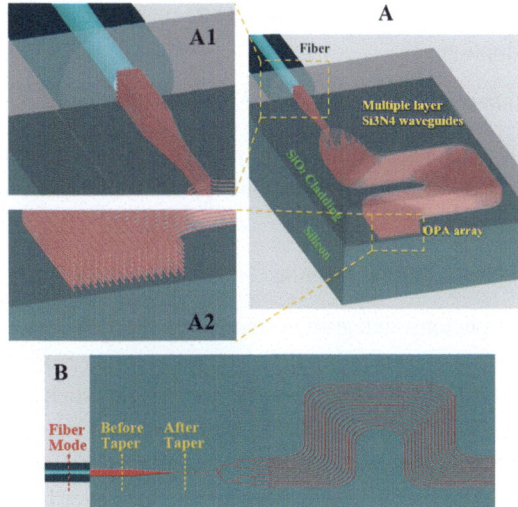

Overview of Solid-State LiDAR

Solid-state LiDAR is an advanced sensing technology that utilizes laser light to detect objects and measure distances, creating detailed 3D maps of environments. Unlike traditional mechanical LiDAR systems that rely on rotating mirrors or spinning components to scan areas, solid-state LiDAR has no moving parts, resulting in significant benefits in durability, reliability, and compactness. The lack of mechanical components reduces wear and tear, making solid-state LiDAR more resilient in challenging environments, which is particularly advantageous for applications in autonomous vehicles, drones, and robotics. Its compact, lightweight design also allows for easier integration into smaller platforms, facilitating the miniaturization of devices without sacrificing performance. Solid-state LiDAR systems can operate at high speeds with minimal latency, providing accurate, real-time depth perception and object detection. This ability to capture precise 3D spatial data reliably and in a compact form factor is making solid-state LiDAR a critical technology in fields requiring high-precision mapping, navigation, and obstacle detection, driving advancements in autonomous navigation, augmented reality, and advanced security systems.

Key Benefits of Solid-State LiDAR

Solid-state LiDAR offers several advantages over traditional mechanical LiDAR systems. The absence of moving parts enhances durability and reliability, reducing the risk of mechanical failure and increasing the lifespan of the system. Miniaturization is another key benefit, as solid-state designs enable more compact and lightweight LiDAR systems that are easier to integrate into various applications, including autonomous vehicles and drones. Additionally, these systems are more cost-efficient to produce and maintain compared to mechanical LiDAR, making them a more practical solution for widespread adoption.

Enhancing Solid-State LiDAR with 3D Photonics

3D photonics significantly improves the performance and miniaturization of solid-state LiDAR systems by enabling advanced integration techniques and innovative photonic components.

High-Resolution Sensing

3D photonic integration allows for the development of high-resolution LiDAR systems capable of generating detailed and accurate 3D maps. Integrated photonic circuits, which incorporate multiple photonic components such as lasers, modulators, and detectors into a single 3D structure, enhance resolution and signal processing capabilities. This integration improves distance measurements and object detection. Additionally, 3D photonics enables the creation of dense sensor arrays that cover larger areas with higher resolution, making LiDAR systems more effective at mapping complex environments.

Improved Range and Accuracy

The use of 3D photonic components extends the range and accuracy of solid-state LiDAR systems, making them suitable for a broader range of applications. High-power laser diodes integrated within a 3D photonic structure increase the system's effective range, allowing it to detect objects at greater distances. Low-loss waveguides further enhance accuracy by reducing signal attenuation, ensuring more reliable and precise distance measurements.

Advanced Signal Processing

Integrating photonic and electronic components within a 3D structure enhances the signal processing capabilities of LiDAR systems. On-chip processing enables real-time data analysis, reducing latency and improving system responsiveness. Additionally, advanced photonic components such as low-noise detectors and optical filters help minimize signal noise, leading to clearer and more accurate LiDAR data.

Applications of Solid-State LiDAR

Solid-state LiDAR systems enhanced by 3D photonics have a wide range of applications across multiple industries, including autonomous vehicles, industrial automation, and aerial mapping.

Autonomous Vehicles

Solid-state LiDAR plays a crucial role in autonomous vehicle navigation by providing high-resolution 3D mapping. These systems enable obstacle detection and avoidance, allowing vehicles to operate safely in complex environments. LiDAR-generated 3D maps also support precise navigation and route planning, ensuring that autonomous vehicles can make informed driving decisions.

Industrial Automation

In industrial settings, solid-state LiDAR is used for automation and safety monitoring. LiDAR-equipped robots can navigate and interact with their environment more accurately, improving efficiency in manufacturing and warehousing. Additionally, LiDAR systems help monitor industrial environments for safety hazards, preventing accidents and ensuring worker protection.

Drones and Aerial Mapping

Drones equipped with solid-state LiDAR systems can capture highly detailed 3D maps from the air, supporting applications such as surveying, urban planning, and environmental monitoring. In agriculture, LiDAR technology helps assess soil conditions, monitor crop health, and optimize irrigation and fertilization practices, improving overall efficiency in precision farming.

Enhancing AI with 3D Photonics

3D photonics also plays a pivotal role in advancing artificial intelligence by enabling the development of more powerful and efficient computing systems.

Neuromorphic Computing

Neuromorphic computing aims to mimic the structure and function of the human brain to achieve high-efficiency data processing. 3D photonic integration enables the development of photonic neuromorphic chips, which use light for data transmission and processing. These chips offer significantly higher processing speeds and lower power consumption compared to traditional electronic neuromorphic chips. Additionally, 3D photonic circuits allow for parallel data processing, further enhancing computational efficiency and AI performance.

Optical Neural Networks

Optical neural networks leverage photonic components to perform neural network operations, offering advantages in speed and energy efficiency. The use of 3D photonic integration allows optical neural networks to process data at the speed of light, greatly surpassing the capabilities of electronic neural networks. Additionally, optical neural networks consume less power, making them ideal for AI applications that require substantial computational power with minimal energy consumption.

AI Accelerators

AI accelerators, designed to optimize AI workloads, benefit from 3D photonics through improved data transmission and thermal management. Integrated photonic circuits enhance AI accelerators by enabling faster and more efficient communication between processing units, reducing latency and improving overall system performance. Additionally, 3D photonic integration allows for innovative cooling solutions, such as microfluidic cooling, which helps manage the heat generated by high-performance AI accelerators.

Applications of AI Enhanced by 3D Photonics

AI systems enhanced by 3D photonics have transformative applications across various industries, from machine learning to real-time data analysis.

Machine Learning

Machine learning algorithms greatly benefit from the high-speed data processing capabilities enabled by 3D photonic systems. The ability to train AI models at an accelerated pace reduces the time required for AI development, allowing for faster advancements in AI applications. Additionally, real-time inference capabilities enabled by photonic systems make AI applications more responsive and efficient.

Real-Time Data Analysis

AI systems integrated with 3D photonics excel in real-time data analysis, providing valuable insights across multiple sectors. In healthcare, AI-driven photonic systems analyze medical data in real-time, assisting in diagnosis, treatment planning, and patient monitoring. In the financial industry, AI algorithms process vast amounts of financial data to detect fraud, assess risks, and make real-time investment decisions.

Autonomous Systems

Autonomous systems, such as robots and drones, benefit from AI enhanced by 3D photonics for improved decision-making and operational efficiency. AI-powered robots use real-time data analysis to navigate and interact with their environment effectively. Drones equipped with AI and 3D photonic systems can autonomously navigate complex environments, performing tasks such as surveying, monitoring, and delivery with greater precision and reliability.

Future Directions

The future of 3D photonics in solid-state LiDAR and AI involves continuous advancements in performance, scalability, and reliability. The integration of quantum photonics with 3D photonic systems could lead to breakthroughs in quantum computing and secure communication. AI-driven design and optimization are expected to further enhance the fabrication and functionality of 3D photonic devices, reducing development time and improving efficiency. Additionally, research into advanced materials with improved thermal, mechanical, and optical properties will continue to expand the capabilities of 3D photonic systems.

Conclusion

The integration of 3D photonics into solid-state LiDAR and AI is revolutionizing these technologies, significantly enhancing their performance and miniaturization. High-resolution sensing, improved range and accuracy, and advanced signal processing are just a few of the benefits that 3D photonics brings to solid-state LiDAR systems. In AI, 3D photonics enables the development of high-speed, energy-efficient computing systems, including neuromorphic chips, optical neural networks, and AI accelerators. By leveraging these innovations, next-generation devices and systems can meet the growing demands of modern sensing, autonomous systems, and artificial intelligence applications.

8.3 Quantum Computing and Sensing: Role of 3D PICs in Advancing Quantum Technologies

Quantum computing and sensing are among the most exciting frontiers in modern science and technology, promising breakthroughs in fields ranging from cryptography and drug discovery to precise environmental monitoring. The development of 3D Photonic Integrated Circuits (PICs) has been pivotal in advancing these quantum technologies by enabling previously unachievable integration density, performance, and scalability. In quantum computing, 3D PICs allow for the integration of multiple quantum photonic components—such as qubits, phase shifters, and waveguides—into a single compact platform, supporting the manipulation and entanglement of quantum states with higher precision and efficiency. This high-density integration minimizes signal loss and crosstalk, critical for building scalable quantum systems capable of complex computations. For quantum sensing, 3D PICs enhance sensitivity and resolution by creating miniaturized systems that can detect minute changes in physical parameters, such as magnetic fields or temperature, at the quantum level. Innovations in material engineering, thermal management, and precision alignment within 3D PICs are driving these advancements, enabling robust quantum systems that are resilient to environmental noise and capable of operating at room temperature. This chapter explores how 3D PICs are transforming the landscape of quantum computing and sensing, detailing the technological innovations that make them essential for next-generation quantum applications (Fig. 8.3).

Overview of Quantum Technologies
Quantum technologies harness the unique principles of quantum mechanics—such as superposition, entanglement, and wave-particle duality—to perform tasks that are beyond the reach of classical systems. Quantum computing relies on quantum bits, or qubits, which can exist in multiple states simultaneously, enabling parallel processing of information and promising exponential speedups for specific types of computations, such as factoring large numbers, optimizing complex systems, and simulating molecular structures. This capability could revolutionize fields like cryptography, materials science, and

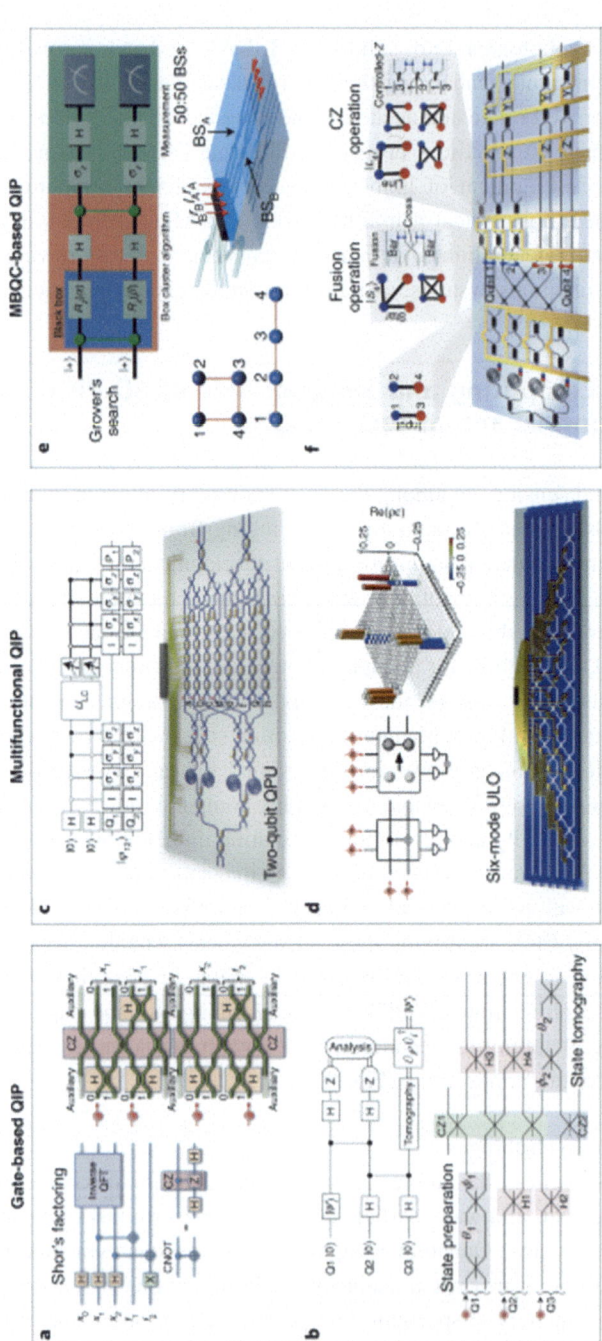

Fig. 8.3 Schematic of a 3D PIC for quantum computing, showing qubits and integrated photonic components for quantum processing. https://www.nature.com/articles/s41566-019-0532-1

8.3 Quantum Computing and Sensing: Role of 3D PICs ...

artificial intelligence by solving problems that are infeasible for even the most powerful classical computers. Quantum sensing, on the other hand, leverages quantum states to achieve extremely sensitive measurements, allowing detection of tiny changes in physical quantities like magnetic fields, time, or temperature. By exploiting quantum entanglement and coherence, quantum sensors can reach sensitivities far beyond classical limits, enabling breakthroughs in medical imaging, environmental monitoring, and navigation. Together, these quantum technologies hold transformative potential, as they push the boundaries of what is scientifically and technologically achievable, opening new possibilities in both scientific research and industrial applications.

Key Benefits of Quantum Technologies

Quantum technologies offer significant advantages over classical systems, revolutionizing computing, sensing, and communication. One of the most notable benefits is exponential speedup, where quantum computing can solve specific problems, such as factoring large numbers and simulating complex molecular structures, much faster than classical computers. Ultra-sensitivity is another critical advantage, as quantum sensors can detect minute changes in physical quantities like magnetic fields with unprecedented precision. Security is also greatly enhanced through quantum communication methods such as quantum key distribution (QKD), which offers theoretically unbreakable encryption.

Role of 3D PICs in Quantum Computing

3D Photonic Integrated Circuits (PICs) are crucial for advancing quantum computing by providing the infrastructure necessary to manipulate and control qubits with high precision and efficiency. These circuits facilitate the compact and scalable integration of quantum components, making quantum computing more practical and accessible.

Integration of Quantum Components

3D PICs enable the seamless integration of essential quantum components, including single-photon sources, waveguides, and detectors, into a single platform. Single-photon sources, such as quantum dots or defect centers in diamond, are integrated with waveguides and other photonic elements, ensuring efficient coupling and manipulation of photons for quantum information processing. High-precision waveguides guide photons with minimal loss and crosstalk, preserving quantum coherence and fidelity. Single-photon detectors embedded in 3D PICs allow for efficient and accurate measurement of quantum states, which is critical for quantum computation and error correction.

Quantum Gates and Circuits

Quantum gates and circuits form the fundamental building blocks of quantum computers, and 3D PICs provide an ideal platform for their implementation. Optical quantum gates, integrated within 3D PICs, manipulate quantum states by enabling operations such as superposition and entanglement, which are essential for quantum computing. The scalability of 3D PICs allows for the dense integration of multiple quantum gates and circuits on a single chip, paving the way for large-scale quantum processors.

Error Correction and Fault Tolerance

Error correction is essential for practical quantum computing, and 3D PICs play a significant role in achieving fault tolerance. These circuits support the integration of error correction codes, which protect quantum information from decoherence and operational errors. The ability to integrate multiple redundant components within a 3D structure enhances fault tolerance, ensuring reliable quantum operations over extended periods.

Role of 3D PICs in Quantum Sensing

Quantum sensing applications benefit immensely from 3D PICs, which enhance the sensitivity, resolution, and integration of quantum sensors, making them more precise and practical.

Enhanced Sensitivity and Resolution

3D PICs improve the sensitivity and resolution of quantum sensors by integrating advanced photonic components and quantum devices. High-precision waveguides guide quantum states with minimal loss and dispersion, leading to more accurate measurements. Integrated interferometers, crucial for high-precision measurements, detect minute phase changes, enabling ultra-sensitive detection capabilities for applications such as medical imaging and gravitational wave observation.

Miniaturization and Portability

The compact and scalable nature of 3D PICs allows for the miniaturization of quantum sensors, making them more portable and practical for real-world applications. Compact designs integrate multiple photonic and quantum components within a single structure, reducing the overall size of quantum sensors. On-chip integration of quantum sensors with electronic readout and control circuits further simplifies system design, reducing complexity and improving usability in diverse environments.

Applications of Quantum Sensing

Quantum sensing technologies enhanced by 3D PICs are transforming fields such as biomedical imaging, geophysics, and environmental monitoring.

- Magnetic Field Sensing: Quantum sensors based on 3D PICs detect extremely weak magnetic fields with high precision, benefiting applications like brain imaging and mineral exploration.

8.3 Quantum Computing and Sensing: Role of 3D PICs ...

- Gravitational Wave Detection: Advanced interferometers integrated into 3D PICs improve the sensitivity of gravitational wave detectors, enabling the observation of astrophysical events.
- Temperature and Pressure Sensing: Quantum sensors measure temperature and pressure with unprecedented accuracy, making them valuable for industrial monitoring and environmental sensing.

Challenges and Innovations

While 3D PICs provide numerous advantages, several challenges must be addressed to fully realize their potential in quantum technologies.

Material Compatibility

Ensuring compatibility between different materials used in 3D PICs is critical for maintaining quantum coherence and device reliability. Thermal expansion mismatches between materials can lead to mechanical stress and alignment issues, necessitating the development of materials with matched thermal properties. Interface quality between photonic and quantum materials is also crucial, as high-quality bonding and epitaxial growth techniques are needed to ensure efficient signal transmission.

Fabrication Complexity

The fabrication of 3D PICs involves highly complex processes that require precise control and alignment. Precision lithography is essential for creating high-accuracy photonic structures and quantum components, while integration techniques such as direct wafer bonding and selective area growth must be optimized for scalable manufacturing.

Thermal Management

Effective thermal management is crucial to prevent overheating and maintain stable quantum operations. Heat dissipation is particularly important for quantum devices that operate at cryogenic temperatures, requiring advanced cooling solutions such as microfluidic cooling integrated with 3D photonic components. Temperature control across the 3D PIC structure is necessary to prevent thermal-induced stresses that could disrupt quantum coherence.

Future Directions

The future of 3D PICs in quantum computing and sensing involves continuous innovation to enhance performance, scalability, and reliability.

- Nanophotonic Structures: Developing nanophotonic structures with tailored optical properties will improve integration density and device performance.
- AI-Driven Design: Artificial intelligence and machine learning can optimize the design and fabrication of 3D PICs, reducing development time and enhancing efficiency.

- Advanced Materials: Research into new materials with improved thermal, mechanical, and optical properties will further enhance the capabilities of 3D photonic systems.

Quantum Internet

The development of a quantum internet, which enables the transmission of quantum information over long distances, will greatly benefit from advancements in 3D PICs. Quantum repeaters integrated into 3D PICs will extend the range of quantum communication networks, ensuring reliable long-distance quantum information transfer. Secure communication using quantum key distribution (QKD) enhanced by 3D PICs will protect data from cyber threats, providing ultra-secure networking solutions.

Hybrid Quantum Systems

Combining different quantum technologies, such as superconducting qubits and photonic qubits, into a single 3D PIC will enable the development of hybrid quantum systems with enhanced capabilities. Interoperability between different quantum technologies will lead to more versatile and powerful quantum systems. Performance optimization will leverage the strengths of different quantum platforms, ensuring that quantum computing and sensing applications reach their full potential. The integration of 3D PICs into quantum computing and sensing is revolutionizing these fields by providing compact, scalable, and high-performance solutions. Quantum computing benefits from 3D PICs through improved quantum gate implementation, error correction, and fault tolerance. In quantum sensing, 3D PICs enhance sensitivity, resolution, and portability, enabling advanced applications in scientific research and industry. While challenges such as material compatibility, fabrication complexity, and thermal management remain, continuous advancements in nanophotonics, AI-driven design, and quantum networking will drive the next generation of quantum technologies. By leveraging these innovations, 3D PICs will play a crucial role in shaping the future of quantum computing, secure communication, and ultra-precise sensing systems.

Conclusion

3D Photonic Integrated Circuits (PICs) are transforming the fields of quantum computing and sensing by enabling the integration of multiple quantum components into compact, scalable platforms, which are essential for achieving high performance and functionality in quantum systems. Through 3D integration, diverse components such as qubits, waveguides, and phase shifters can be layered vertically, maximizing space efficiency and minimizing signal loss, which is crucial for maintaining quantum coherence and precision. This dense integration also supports more complex quantum operations and parallel processing, which enhance computational power and measurement sensitivity. Innovations in integration techniques—such as wafer bonding and through-silicon vias—allow for seamless stacking of materials that are often challenging to combine, like silicon, indium phosphide, and superconducting materials. Additionally, advanced thermal management solutions in 3D PICs, such as microfluidic cooling and heat spreaders, help to dissipate heat generated by densely packed quantum components, ensuring stability and reliability. As research progresses, 3D PICs will be instrumental in driving breakthroughs in quantum applications by providing the foundational architecture needed for high-speed, energy-efficient, and miniaturized quantum systems. These advancements are expected to revolutionize scientific research, cryptography, precision sensing, and secure communication, establishing 3D PICs as a cornerstone in the rapidly evolving quantum technology landscape.

Modeling and Simulation Tools

9

- **Electronic Design Automation (EDA) for Photonics**: Adapting traditional EDA tools for photonic circuits.
- **Advanced Simulation Techniques**: Modeling interactions between various layers and components.
- **Predictive Modeling of 3D Photonic Structures**: Ensuring optimal performance and reliability.

9.1 Electronic Design Automation (EDA) for Photonics: Adapting Traditional EDA Tools for Photonic Circuits

The development of photonic integrated circuits (PICs) has driven the need for specialized design methodologies and tools that go beyond traditional Electronic Design Automation (EDA) developed for electronic circuits. EDA for photonics involves adapting these conventional tools to address the unique characteristics of photonic components, such as waveguides, modulators, and lasers, which rely on light manipulation rather than electric current. Traditional EDA tools, which focus on electronic parameters like voltage, current, and resistance, must be modified to account for photonic-specific factors like wavelength, phase, polarization, and optical loss. This requires new modeling techniques and simulation tools capable of handling both optical and electronic behaviors, as well as co-design methodologies that can simulate the interaction between photonic and electronic elements within hybrid integrated circuits. Developing these tools presents challenges, including creating accurate models for photonic devices, ensuring compatibility between photonic and electronic design processes, and managing thermal effects in densely packed photonic circuits. Despite these challenges, the advancement of photonic EDA tools offers significant benefits, enabling more precise designs, faster prototyping, and optimized performance for PICs. This chapter delves into these EDA adaptations, examining the technical challenges, the innovative solutions being developed, and how they are laying the groundwork for the efficient design and large-scale manufacturing of photonic circuits, which are essential for applications in data communications, sensing, and computing.

Overview of EDA Tools

Electronic Design Automation (EDA) tools are specialized software applications that enable engineers to design, simulate, analyze, and verify electronic circuits with high precision. These tools automate many stages of the design process, including schematic capture, layout design, and circuit simulation, significantly reducing the time and cost associated with developing complex electronic systems like microprocessors, communication devices, and integrated circuits. EDA tools streamline workflows by allowing engineers to model circuit behavior before physical fabrication, detect errors early, optimize component placement, and ensure the circuit meets performance specifications. They encompass various software suites tailored for specific tasks, such as circuit simulation (e.g., SPICE), layout design (e.g., Cadence Virtuoso), and timing analysis (e.g., Synopsys PrimeTime). Additionally, EDA tools support design rule checks (DRC) and layout versus schematic (LVS) verification to ensure that the design adheres to manufacturing constraints and matches the intended schematic. With advancements in technology, EDA tools have evolved to handle increasingly complex designs, enabling the development of smaller, faster, and more efficient electronic systems across industries such as computing, telecommunications, and automotive electronics. As a result, EDA has become an indispensable resource in modern electronics, driving innovation and accelerating the pace of technology development (Fig. 9.1).

Key Functions of Traditional EDA Tools

Electronic Design Automation (EDA) tools play a vital role in circuit design by streamlining processes from schematic creation to verification. Schematic capture enables designers to create circuit diagrams that illustrate electrical connections between components. Simulation provides a virtual testing environment to analyze circuit behavior under various conditions, ensuring optimal performance. Layout design facilitates the physical arrangement of components on a chip, ensuring proper connectivity and functionality. Verification ensures that the design meets specified requirements and performs as intended by conducting rigorous checks and tests.

Adapting EDA Tools for Photonic Circuits

The adaptation of traditional EDA tools for photonic circuits requires addressing the distinct characteristics of photonics, such as light propagation, optical material properties,

9.1 Electronic Design Automation (EDA) for Photonics: Adapting ...

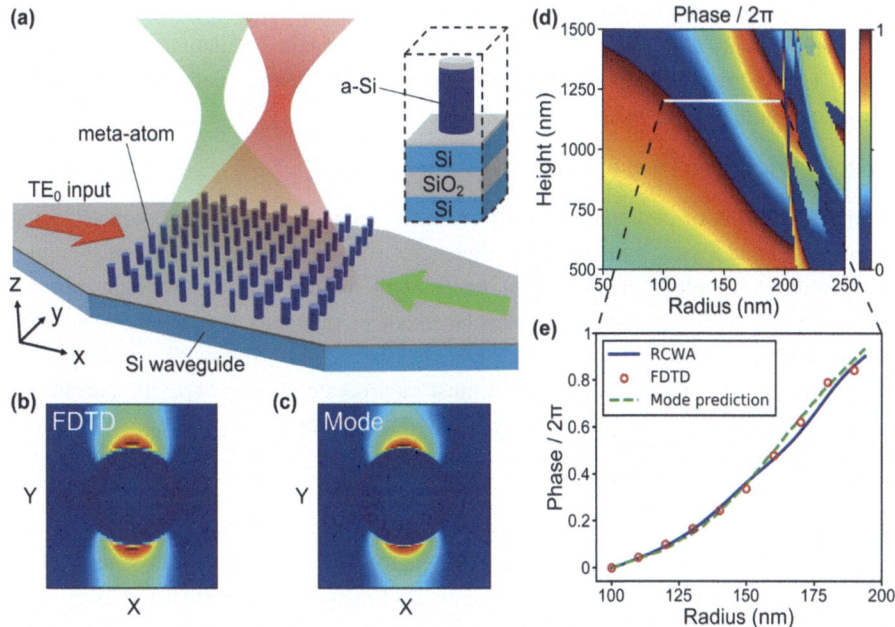

Fig. 9.1 Diagram of the EDA workflow adapted for photonic circuits, from design to simulation. https://www.degruyter.com/document/doi/10.1515/nanoph-2022-0344/html

and wave interference. Unlike electronic circuits, which rely on voltage and current, photonic circuits manipulate light waves, necessitating modifications to existing tools and the development of specialized functionalities tailored for photonic design.

Schematic Capture for Photonic Circuits

Schematic capture tools for photonic circuits must accommodate both electronic and photonic components, enabling seamless integration of hybrid systems. Hybrid components such as electro optic modulators, photodetectors, and waveguides must be accurately represented with both electrical and optical connections. The development of comprehensive component libraries is essential to facilitate efficient design, incorporating models for photonic devices like lasers, modulators, detectors, and passive elements such as waveguides and couplers.

Simulation of Photonic Circuits

Simulating photonic circuits requires specialized tools that can accurately model the behavior of light and its interaction with different materials. Optical simulation engines must complement traditional electronic simulators to handle phenomena like wave propagation, interference, and diffraction. Techniques such as finite-difference time-domain

(FDTD), beam propagation method (BPM), and eigenmode expansion (EME) are essential for precise optical modeling. Additionally, mixed-domain simulation capabilities are required to analyze interactions between electronic and photonic components, ensuring accurate co-simulation of hybrid systems.

Layout Design for Photonic Circuits

The physical layout of photonic circuits introduces unique considerations such as waveguide routing, bend losses, and coupling efficiencies. Waveguide layout tools must support precise routing, enabling the design of bends with minimal signal loss and appropriate spacing to avoid crosstalk. Accurate placement of photonic components—including gratings, mirrors, and splitters—is crucial to optimizing performance. To ensure manufacturability, design rule checking (DRC) must incorporate photonic-specific constraints, such as minimum waveguide bend radii and alignment tolerances.

Verification of Photonic Circuits

Verification ensures that photonic circuits meet design specifications and perform reliably under varying conditions. Optical performance metrics such as insertion loss, return loss, and crosstalk must be evaluated to guarantee circuit functionality. Thermal and mechanical analysis is essential, as photonic circuits are highly sensitive to temperature variations and mechanical stress. Additionally, fabrication tolerance analysis must account for potential process variations, ensuring that the design remains robust under real-world manufacturing conditions.

Challenges in Adapting EDA Tools for Photonics

Despite the significant advantages of adapting EDA tools for photonics, several challenges must be addressed due to fundamental differences between electronic and photonic systems.

Modeling and Simulation Accuracy

Accurately modeling light behavior and its interaction with materials is computationally demanding. Complex optical phenomena such as interference, diffraction, and nonlinear effects require sophisticated simulation techniques that balance accuracy and computational efficiency. Furthermore, multi-scale modeling is necessary to accommodate photonic components that range from nanometer-scale waveguides to millimeter-scale fibers, demanding highly versatile simulation tools.

Integration of Photonic and Electronic Design

To develop hybrid photonic-electronic circuits, seamless integration between electronic and photonic design workflows is essential. Mixed-domain design tools must support co-design and co-simulation of electronic and photonic components, ensuring compatibility between the two domains. Additionally, data exchange and interoperability between different design tools and simulation engines must be optimized to enable efficient cross-domain collaboration.

Fabrication and Manufacturing Constraints

Designing photonic circuits that can be reliably manufactured requires consideration of various fabrication constraints. Process variability in photonic fabrication can significantly impact device performance, necessitating tools that incorporate design margins for robustness. Design for manufacturability (DFM) principles should be embedded into EDA tools to ensure high-yield fabrication and scalability.

Benefits of EDA Tools for Photonics

The adaptation of EDA tools for photonic circuits offers significant advantages, enhancing design efficiency, performance, and innovation.

Increased Design Efficiency

EDA tools streamline the design process by automating many tasks, reducing development time and effort. Automated layout and routing tools for waveguides enhance design productivity, while parametric design capabilities enable designers to explore multiple configurations and optimize performance quickly.

Improved Performance and Reliability

EDA tools facilitate precise modeling and simulation, ensuring that photonic circuits meet performance requirements. Accurate simulation engines help identify potential performance issues early in the design cycle, while verification and testing tools ensure compliance with specifications, reducing the likelihood of failures in real-world applications.

Accelerated Innovation

By enabling rapid prototyping and iteration, EDA tools accelerate the development of new photonic technologies. Prototyping and iteration capabilities allow researchers and engineers to refine designs efficiently, while collaborative design environments support interdisciplinary teamwork, fostering innovation across multiple domains.

Future Directions

The future of EDA for photonics will see continuous advancements in modeling, simulation, and integration to meet the evolving demands of photonic technology.

AI and Machine Learning

Incorporating artificial intelligence and machine learning into EDA tools will enhance automation and optimization. AI-driven design optimization can refine design parameters for performance, manufacturability, and yield, reducing development cycles. Predictive modeling using machine learning can anticipate process variations and environmental influences, enabling more robust photonic designs.

Advanced Simulation Techniques

Developing more sophisticated simulation techniques will be critical for handling the complexity and scale of modern photonic circuits. Quantum simulation tools will be necessary as quantum photonic technologies advance, requiring models based on quantum mechanics to accurately predict circuit behavior. Multi-physics simulation, incorporating thermal, mechanical, and optical effects, will ensure comprehensive analysis of photonic circuits under real-world conditions.

Integration with Fabrication Technologies

Ensuring that EDA tools are closely aligned with fabrication processes will improve design-to-manufacturing workflows. Process design kits (PDKs) tailored for photonic technologies will optimize designs for specific fabrication processes, enhancing yield and performance. Additionally, feedback and iteration mechanisms will allow designers to refine photonic circuits based on real fabrication data, continuously improving manufacturing outcomes.

Conclusion

Adapting traditional EDA tools for photonic circuits is a crucial step toward advancing photonic technology. By incorporating specialized functionalities tailored to the unique properties of light, these tools enable efficient and accurate design, simulation, and verification of photonic circuits. The benefits of EDA for photonics include increased design efficiency, improved performance, and accelerated innovation, making it a key enabler for emerging applications in telecommunications, data centers, and advanced computing. As research and development continue, the future of EDA for photonics will be shaped by AI-driven design, advanced simulation techniques, and close integration with fabrication technologies, driving the evolution of next-generation photonic systems.

9.2 Advanced Simulation Techniques: Modeling Interactions Between Various Layers and Components

The complexity of modern photonic integrated circuits (PICs) demands advanced simulation techniques to model the intricate interactions between various layers and components accurately. Unlike electronic circuits, where current flows in a predictable manner, PICs involve the manipulation of light, which requires precise modeling of optical phenomena such as reflection, refraction, interference, and diffraction. Advanced simulation techniques—such as finite-difference time-domain (FDTD), beam propagation method (BPM), and finite element method (FEM)—are essential for predicting how light behaves as it

9.2 Advanced Simulation Techniques: Modeling Interactions ...

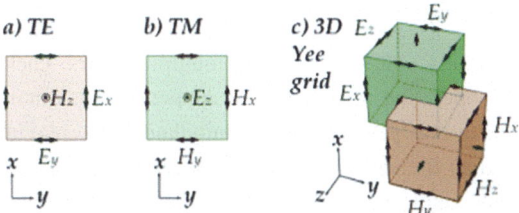

Fig. 9.2 Example of FDTD (Finite-Difference Time-Domain) simulation results showing light propagation in a 3D structure. https://en.wikipedia.org/wiki/Finite-difference_time-domain_method

interacts with different materials and structures in multi-layered photonic devices. These simulations allow engineers to analyze how each layer influences the overall performance, identify potential losses or signal distortions, and optimize the design to achieve higher efficiency and functionality. Additionally, thermal and mechanical simulations are integrated to assess the effects of temperature changes and mechanical stress, which can impact alignment and performance in densely packed 3D PICs. This chapter delves into these advanced simulation methodologies, highlighting the key challenges such as computational intensity and accuracy requirements, as well as the benefits these techniques offer in ensuring reliable and optimized PIC designs for applications in telecommunications, data centers, and sensing technologies (Fig. 9.2).

Overview of Photonic Integrated Circuits

Photonic integrated circuits (PICs) bring together multiple photonic components—such as waveguides, modulators, detectors, and lasers—onto a single chip, creating highly compact and efficient devices for manipulating light. The overall performance of these circuits relies on the precise interactions between different layers and components, which must be carefully designed and controlled. For instance, waveguides must efficiently direct light between different components with minimal loss, while modulators require precise alignment to adjust the properties of the light passing through. The integration of detectors and lasers further adds complexity, as each component must operate in harmony without causing unwanted reflections, interference, or signal distortions. Additionally, materials with different refractive indices are often stacked in layers, affecting how light propagates within the circuit and introducing challenges in managing signal integrity and thermal stability. The interaction between these components in PICs is highly sensitive to even minor variations in fabrication and alignment, making precise modeling, simulation, and optimization essential to ensure reliable performance. This level of integration enables PICs to achieve high data rates, low power consumption, and compact form factors, making them indispensable in telecommunications, data processing, and sensing applications.

Key Layers in Photonic Integrated Circuits (PICs)

Photonic Integrated Circuits (PICs) consist of multiple layers, each serving a distinct function in the overall operation of the device. Active layers contain components such as lasers and modulators that generate and manipulate light, playing a crucial role in signal processing and data transmission. Passive layers include waveguides and splitters that guide and distribute light, ensuring efficient signal routing with minimal loss. The substrate provides structural support and may also incorporate integrated electronic components for control and signal processing, enhancing the functionality and reliability of the PIC.

Importance of Accurate Modeling

Accurate modeling of interactions between layers and components is essential for optimizing performance, reliability, and manufacturability. Design optimization ensures that potential issues are identified early in the design process, preventing costly revisions. Reliability modeling predicts how factors such as thermal effects and material properties impact device longevity, allowing for proactive design improvements. Manufacturability considerations ensure that PIC designs can be produced with high yield and consistent quality, minimizing defects and performance variations.

Challenges in Modeling Interactions

Modeling interactions in PICs presents several challenges due to the complex behavior of light and its interaction with various materials and structures.

Photonic circuits operate across multiple scales, requiring precise modeling techniques to account for these variations. Nanophotonic structures involve interactions at the nanoscale, where quantum and optical effects must be accurately captured. Macroscale integration deals with the overall layout and routing of waveguides and components, ensuring proper signal transmission across the entire PIC.

The optical properties of materials significantly impact PIC performance, requiring accurate modeling of factors such as refractive index and absorption. Refractive index variations due to temperature fluctuations, fabrication tolerances, and wavelength dependency can affect device behavior and signal integrity. Material absorption must be minimized to reduce signal loss and improve efficiency, making it critical to select and model materials accurately.

Thermal and mechanical stresses can influence the alignment and functionality of photonic components. Thermal expansion of different materials at varying rates can cause misalignment, leading to performance degradation. Mechanical stresses introduced during fabrication and packaging can impact structural integrity and device reliability, necessitating careful modeling and mitigation strategies.

Advanced Simulation Techniques

To address these challenges, advanced simulation techniques are employed to model the interactions between different layers and components in PICs.

The Finite-Difference Time-Domain (FDTD) method is a widely used numerical technique for solving Maxwell's equations in the time domain, enabling detailed simulation of electromagnetic wave propagation in complex photonic structures. This method discretizes Maxwell's equations in time and space, allowing the simulation of broadband optical behavior and transient phenomena. Yee's lattice provides a staggered grid structure where electric and magnetic field components are updated iteratively, ensuring accurate modeling of wave propagation in nanophotonic structures. FDTD is well-suited for modeling nanoscale photonic devices, capturing intricate material interfaces and multilayered PIC structures with high spatial and temporal resolution. Operating in the time domain allows FDTD to simultaneously capture responses over a wide range of frequencies, making it ideal for designing photonic components like waveguides, modulators, and resonators.

The Beam Propagation Method (BPM) efficiently models the propagation of optical beams through waveguides and photonic structures. The paraxial approximation simplifies simulations by assuming primarily forward-propagating waves, making BPM computationally efficient. This method enables the simulation of large-scale photonic circuits with reduced computational resources compared to full-wave methods like FDTD. BPM is particularly effective for evaluating mode propagation, bend losses, and coupling effects in photonic waveguides.

The Eigenmode Expansion (EME) method provides a frequency-domain approach for analyzing the modal properties of photonic structures. Modal decomposition simplifies analysis by expressing optical fields as eigenmodes, improving computational efficiency. Frequency-domain analysis offers detailed insights into dispersion, resonances, and waveguide coupling behavior. Coupled-mode theory facilitates the study of interactions between waveguide modes, optimizing devices such as directional couplers and ring resonators.

Modern photonic designs require simulation tools that integrate multiple physical effects beyond optical interactions. Multi-physics simulation involves simultaneous modeling of optical, thermal, and mechanical effects to predict temperature-induced changes in refractive index and material properties. Mechanical stress analysis evaluates the impact of fabrication and packaging processes on device integrity. Integrated simulation frameworks combine optical, thermal, and mechanical modeling, providing a holistic understanding of device behavior.

Integration with EDA Tools

To streamline the design and verification of PICs, advanced simulation techniques are integrated with Electronic Design Automation (EDA) tools.

Simulation tools must seamlessly integrate into the overall photonic design flow. Schematic capture allows designers to define photonic circuits using an intuitive graphical interface, which is then linked to simulation tools for validation. Layout design tools ensure that the physical arrangement of photonic waveguides and components aligns with manufacturability constraints.

Simulation tools enable parameter sweep and optimization, allowing designers to explore different design configurations efficiently. Design space exploration helps identify optimal trade-offs between performance, efficiency, and fabrication constraints. Optimization algorithms, such as genetic algorithms and machine learning techniques, automate the refinement of photonic circuit designs.

Comprehensive verification ensures that photonic circuits meet performance and reliability standards. Performance metrics evaluation provides insights into insertion loss, return loss, crosstalk, and bandwidth. Robustness analysis assesses design tolerance to fabrication variations, environmental conditions, and operational fluctuations.

Case Studies and Applications

Advanced simulation techniques are applied across various photonic applications, demonstrating their effectiveness in optimizing PIC designs.

Simulation tools aid in the design and optimization of high-speed modulators for efficient electro-optic conversion. Waveguide design optimization ensures minimal loss and efficient light modulation. Thermal management analysis evaluates temperature effects, guiding the development of cooling strategies for stable operation.

Advanced simulations enhance the performance and reliability of optical interconnects used in data centers and high-performance computing. Mode coupling analysis improves coupling efficiency between optical fibers and waveguides, minimizing signal loss. Mechanical stability modeling predicts the impact of stress on interconnect performance, ensuring long-term reliability.

Simulation tools play a vital role in developing quantum photonic circuits for secure communication and quantum computing. Single-photon source optimization ensures high efficiency and minimal loss for quantum information processing. Quantum gate analysis refines the design of high-fidelity quantum operations within photonic circuits.

Future Directions

The evolution of simulation techniques in photonics is driven by advancements in AI, quantum mechanics, and integrated simulation environments.

Incorporating AI into simulation tools enhances design automation and predictive modeling. AI-driven optimization accelerates the identification of optimal design parameters. Predictive modeling anticipates performance variations based on simulation data, improving robustness.

As quantum photonic technologies progress, simulation tools must incorporate quantum mechanical models. Quantum waveguide simulations predict photon behavior in quantum circuits. Entanglement modeling enhances the understanding of quantum information transfer.

Developing unified platforms that integrate optical, thermal, mechanical, and quantum simulations will revolutionize photonic design. Holistic design approaches ensure comprehensive analysis of PICs across multiple domains. Interdisciplinary collaboration tools facilitate joint development between optical, electrical, and mechanical engineers.

Conclusion

Advanced simulation techniques are essential for accurately modeling interactions in PICs, ensuring optimal performance, efficiency, and manufacturability. Techniques such as FDTD, BPM, and EME provide powerful tools for analyzing optical behavior, while multi-physics simulation integrates thermal and mechanical effects. By integrating these simulations with EDA tools, designers can streamline workflows, optimize performance, and push the boundaries of photonic technology. Future developments in AI, quantum simulation, and integrated environments will further enhance these capabilities, driving the next generation of photonic innovation.

9.3 Predictive Modeling of 3D Photonic Structures: Ensuring Optimal Performance and Reliability

Introduction

In the rapidly advancing field of photonics, the development and optimization of three-dimensional (3D) photonic structures are crucial for innovations in telecommunications, computing, sensing, and more. These 3D structures enable high integration densities and complex functionalities, which are essential for meeting the growing demands of high-speed data transmission, efficient computation, and precise sensing. However, predicting their performance is challenging due to the intricate interplay of material properties, geometric configurations, and environmental factors such as temperature and humidity. Predictive modeling has become an invaluable tool for ensuring the optimal performance and reliability of these complex photonic systems, allowing engineers to evaluate and refine designs before fabrication. This modeling employs a range of techniques, from analytical methods that provide quick insights into simple structures, to advanced computational simulations like the Finite-Difference Time-Domain (FDTD) and Finite Element Method (FEM) techniques. These simulations allow for detailed analysis of electromagnetic wave behavior, material interactions, and structural resilience under various operating conditions. By accurately forecasting the behavior and limitations of 3D photonic structures, predictive modeling aids in minimizing design iterations, reducing development costs, and accelerating the deployment of advanced photonic technologies in real-world applications.

Importance of Predictive Modeling

The importance of predictive modeling in the development of 3D photonic structures lies in its ability to offer deep insights into the physical phenomena that govern these systems' behavior. By accurately simulating how complex photonic structures interact with light, predictive modeling allows researchers and engineers to optimize designs for maximum efficiency, minimal losses, and enhanced structural reliability before moving to physical prototyping. This is especially valuable in applications such as telecommunications, sensing, and quantum computing, where precise control over light propagation is crucial. Predictive modeling helps designers fine-tune aspects like waveguide configurations, material interfaces, and layer alignment, achieving desired performance outcomes while identifying potential issues—such as signal loss, thermal effects, or structural weaknesses—early in the design phase. This proactive approach not only reduces the time and cost associated with multiple rounds of experimental prototyping and testing but also shortens the product development cycle and increases the reliability of the final device. As a result, predictive modeling is a powerful tool that accelerates innovation and allows for the scalable production of advanced 3D photonic technologies across various high-tech industries (Fig. 9.3).

Techniques for Predictive Modeling

Predictive modeling of 3D photonic structures involves a variety of techniques, each offering different levels of precision and computational complexity. One of the most commonly

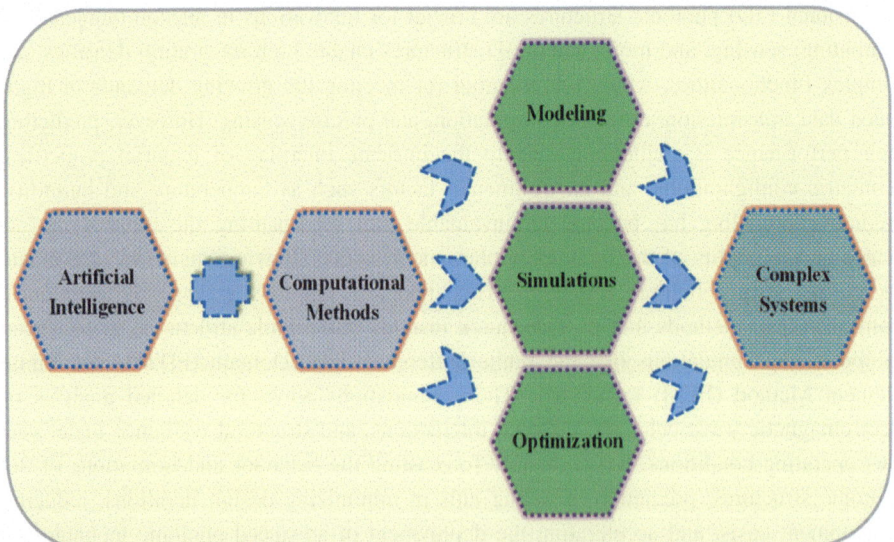

Fig. 9.3 Diagram of predictive modeling used to forecast the behavior and reliability of 3D photonic structures. https://www.mdpi.com/1996-1944/17/14/3521

9.3 Predictive Modeling of 3D Photonic Structures: Ensuring ...

used methods is the Finite-Difference Time-Domain (FDTD) method, a numerical analysis technique that solves Maxwell's equations in both time and space. This method provides detailed insights into the electromagnetic field distributions within photonic structures, making it particularly useful for modeling complex geometries and inhomogeneous materials. Another essential technique is the Beam Propagation Method (BPM), which simulates the propagation of light through optical waveguides and other photonic structures. BPM simplifies the problem by assuming that the light field changes slowly along the direction of propagation, making it computationally less intensive than FDTD.

The Finite Element Method (FEM) is another widely used approach, dividing the photonic structure into a mesh of smaller, simpler elements to calculate electromagnetic fields within each element. This method is highly versatile and capable of handling complex boundaries and material properties. For periodic structures such as photonic crystals and diffraction gratings, Rigorous Coupled-Wave Analysis (RCWA) is particularly effective. This method decomposes electromagnetic fields into Fourier series and solves the resulting set of coupled wave equations, enabling precise analysis of periodic photonic designs.

Recent advancements in machine learning and artificial intelligence (AI) have introduced new possibilities for predictive modeling. AI-driven techniques can identify patterns and correlations in large datasets, allowing for highly accurate predictions of photonic structure performance. These methods enhance traditional modeling approaches by accelerating design optimizations and improving the accuracy of performance forecasts.

Ensuring Optimal Performance

To ensure the optimal performance of 3D photonic structures, predictive modeling must account for various factors, including material properties, geometric configurations, and environmental conditions. Material properties such as refractive index, absorption coefficient, and dispersion must be accurately modeled, as variations in these properties can significantly impact performance, leading to changes in efficiency and reliability. Geometric configurations play a crucial role in how light propagates through a structure. The shape, size, and arrangement of components must be captured with precision in predictive models to ensure the intended performance is achieved.

Environmental conditions such as temperature fluctuations, mechanical stresses, and electromagnetic interference can affect the stability and efficiency of photonic structures. Predictive models must incorporate these factors to ensure reliability under real-world conditions. Additionally, boundary conditions at interfaces and surrounding environments significantly impact overall performance. Accurate modeling of these interactions is essential to achieving optimal functionality.

Optimization techniques further refine predictive modeling. Optimization algorithms, including genetic algorithms, particle swarm optimization, and gradient-based methods, are often used alongside predictive models to identify the best design parameters for achieving desired performance metrics. These techniques enable efficient exploration of design spaces and improve overall system robustness.

Enhancing Reliability

Reliability is a critical aspect of 3D photonic structures, particularly in applications where consistent performance over extended periods is essential. Predictive modeling plays a crucial role in ensuring reliability by analyzing failure modes, assessing manufacturing tolerances, and simulating environmental stresses. Failure mode analysis allows predictive models to identify potential weaknesses in the design and suggest improvements to mitigate risks. Tolerance analysis evaluates the sensitivity of photonic structures to variations in manufacturing processes and material properties, helping to establish acceptable tolerance levels and ensure consistent performance.

Simulating the effects of environmental stressors, such as temperature cycling, humidity, and mechanical vibrations, enables the evaluation of structural durability and robustness under real-world conditions. Additionally, predictive models can estimate the operational lifetime of photonic structures by considering material degradation, thermal effects, and mechanical wear. These insights are valuable for determining maintenance schedules and replacement timelines, ensuring long-term reliability.

Case Studies

Predictive modeling has been instrumental in the design and optimization of photonic crystal fibers (PCFs). By accurately simulating the propagation of light through the periodic microstructure of PCFs, researchers have developed fibers with tailored dispersion properties, high nonlinearity, and low losses. These advancements have led to applications in telecommunications and nonlinear optics, where precise light manipulation is required.

The development of integrated photonic circuits, which combine multiple photonic components on a single chip, has also greatly benefited from predictive modeling. Techniques such as FDTD and FEM have enabled the precise design of waveguides, modulators, and detectors, ensuring optimal performance while minimizing crosstalk and signal degradation. The ability to simulate interactions between photonic and electronic components has been critical in advancing compact and high-efficiency photonic chips.

Another area where predictive modeling plays a vital role is in the design of metamaterials, which exhibit unique electromagnetic properties not found in natural materials. By simulating the interaction of light with subwavelength structures, researchers have been able to create devices with negative refractive indices, perfect lenses, and even cloaking capabilities. These groundbreaking advancements in metamaterial design would not have been possible without sophisticated predictive modeling techniques that account for complex wave interactions and material behaviors. Predictive modeling is essential for designing and optimizing 3D photonic structures, ensuring their efficiency, reliability, and manufacturability. Techniques such as FDTD, BPM, FEM, and RCWA provide powerful tools for accurately simulating light propagation, wave interactions, and material effects. The integration of machine learning and AI further enhances predictive capabilities, enabling faster optimizations and improved accuracy. By accounting for material properties, geometric configurations, environmental conditions, and boundary interactions, predictive models help engineers refine photonic designs and enhance performance. These modeling techniques play a crucial role in advancing photonic technologies, enabling breakthroughs in telecommunications, optical computing, and quantum photonics. As research continues, predictive modeling will remain a cornerstone of photonic innovation, driving the development of next-generation photonic devices.

Predictive modeling is essential for the optimal performance and reliability of 3D photonic structures in advanced technologies. By employing advanced computational techniques—such as finite-difference time-domain (FDTD) and finite element method (FEM) simulations—predictive models account for factors like material properties, geometric configurations, and environmental conditions that impact device behavior. These models enable engineers to analyze how light interacts within intricate 3D structures, helping to identify and mitigate potential issues such as optical loss, signal distortion, or heat buildup before actual fabrication. This modeling approach provides valuable insights that guide the design and optimization process, allowing adjustments to be made to improve efficiency, signal integrity, and robustness. As photonic technologies evolve, the role of predictive modeling will only become more significant, driving the development of innovative, high-performance photonic devices for a wide array of applications, from high-speed data communications to precise sensing and quantum computing. By reducing reliance on physical prototypes and costly testing iterations, predictive modeling accelerates development cycles, supports scalability, and ensures that new devices meet the rigorous demands of modern applications.

Industry Trends and Future Directions 10

- **Emerging Technologies**: Latest advancements and future possibilities in 3D photonics.
- **Market Trends and Industry Outlook**: Analysis of current market dynamics and future growth prospects.
- **Potential Breakthroughs**: Areas with the highest potential for innovation and development.

10.1 Emerging Technologies: Latest Advancements and Future Possibilities in 3D Photonics

Introduction

The field of 3D photonics is advancing rapidly, driven by the demand for photonic devices that are not only faster and more efficient but also highly compact and capable of supporting new applications. Innovations in 3D integration techniques—such as stacking multiple layers of photonic components, creating complex waveguide architectures, and employing advanced materials like silicon photonics and lithium niobate—are pushing the boundaries of what photonic devices can achieve. These advancements are particularly impactful in fields like telecommunications, where high data throughput and minimal latency are crucial, and computing, where photonic processors and interconnects promise to revolutionize data processing speeds and energy efficiency. Additionally, 3D photonics is opening up new possibilities in sensing and imaging, enabling highly sensitive sensors for biomedical and environmental monitoring as well as compact, high-resolution imaging systems for autonomous vehicles. This section explores the latest breakthroughs—such as the development of low-loss vertical interconnections, multi-layer waveguides, and integration with

Photonics worldwide market size

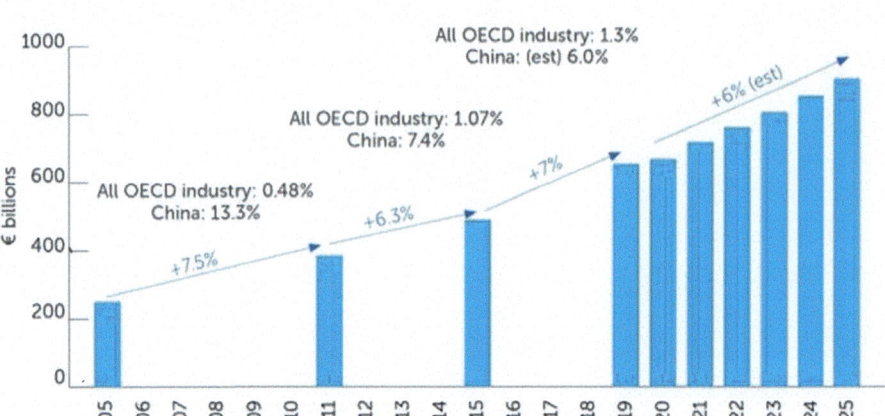

Fig. 10.1 Graph of market growth trends in 3D photonics across key industries. https://www.laserfocusworld.com/executive-forum/article/14305683/multiple-forecasts-project-significant-growth-in-photonics

electronic circuits—that are laying the groundwork for the next generation of photonic devices. As researchers continue to overcome challenges related to thermal management, alignment precision, and fabrication complexity, the future of 3D photonics holds transformative potential for numerous industries, paving the way for revolutionary applications in quantum computing, augmented reality, and high-performance computing (Fig. 10.1).

Latest Advancements in 3D Photonics

3D Photonic Crystals have seen significant advancements, with improved fabrication techniques such as direct laser writing and two-photon polymerization enabling greater precision and complexity. These structures, which feature periodic variations in refractive index, create band gaps that influence light propagation. They are increasingly being used in the development of efficient light sources, such as LEDs and lasers, and in optical fibers for enhanced data transmission.

Integrated Photonics continues to be a major focus area, particularly in silicon photonics, which leverages semiconductor manufacturing processes to create compact and cost-effective photonic devices. This technology is being widely adopted in data centers for high-speed data transfer, in telecommunications for signal processing, and in quantum computing for manipulating qubits.

Metamaterials and Metasurfaces are engineered to exhibit optical properties not found in nature, such as negative refractive indices. Advances in nanofabrication techniques have enabled the creation of 3D metamaterials with tailored optical properties. These materials are being used in applications such as superlenses that surpass the diffraction limit, invisibility cloaks, and devices for controlling light propagation at the nanoscale.

3D Photonic Printing is making significant strides with the adaptation of additive manufacturing techniques for photonic applications. High-resolution 3D printing technologies now allow for the fabrication of complex photonic structures with intricate geometries. These advancements have led to the development of 3D-printed photonic devices, including waveguides, sensors, and micro-optical components, which provide customized solutions for medical imaging, environmental monitoring, and telecommunications.

Nonlinear Photonics is being harnessed in 3D photonic structures to create novel devices that exploit effects such as frequency conversion and self-phase modulation. Advances in material science have led to the discovery of new nonlinear materials with enhanced properties, enabling applications in wavelength conversion for optical communication, all-optical signal processing, and the generation of entangled photons for quantum communication.

Future Possibilities in 3D Photonics

Quantum Photonics is set to revolutionize secure communication, ultra-sensitive sensors, and quantum computing. Researchers are working on developing quantum photonic circuits that manipulate and detect single photons with high precision. However, achieving scalable and fault-tolerant quantum photonic systems remains a challenge, requiring breakthroughs in material science, fabrication techniques, and error correction algorithms.

Topological Photonics is an emerging field exploring topological insulators, which feature robust edge states that are immune to defects and disorder. These materials could enable photonic devices with unprecedented robustness and efficiency, leading to applications such as fault-tolerant photonic circuits, robust optical communication systems, and novel light sources.

Flexible and Wearable Photonics is advancing through developments in flexible and stretchable materials, enabling wearable photonic devices that conform to different shapes and surfaces. These innovations are particularly promising for medical diagnostics, health monitoring, and wearable displays. Potential applications include smart contact lenses with integrated displays, skin-mounted sensors for real-time health monitoring, and flexible photonic circuits for adaptive optics.

Artificial Intelligence in Photonics is unlocking new possibilities for autonomous design, optimization, and control of photonic devices. AI algorithms can analyze vast datasets to identify optimal photonic structures and predict their performance. This could lead to self-optimizing optical networks, intelligent imaging systems, and adaptive photonic devices capable of responding dynamically to environmental changes.

Biophotonic Applications are expanding rapidly, with photonics playing an increasingly critical role in biology and medicine. Future developments could include advanced imaging techniques such as super-resolution microscopy and optogenetics for studying biological processes at the cellular and molecular levels. These technologies could revolutionize medical diagnostics through non-invasive imaging of tissues, real-time monitoring of biochemical processes, and targeted therapies using light-activated drugs.

Impact on Various Industries

In telecommunications, the deployment of integrated photonic circuits and high-speed optical fibers is revolutionizing data transmission. These technologies enable faster speeds, higher bandwidth, and lower latency, supporting the growing demand for data in the 5G era and beyond. Future trends include the development of all-optical networks, where data processing occurs entirely in the optical domain, eliminating electronic conversions and significantly improving network efficiency.

In computing, photonic computing is emerging as a powerful alternative to traditional electronic systems. Integrated photonic circuits and optical interconnects are addressing the limitations of electronic computing in terms of speed and energy efficiency. The future of computing could see the rise of hybrid systems that combine photonic and electronic components, leveraging the strengths of both technologies for superior performance and reduced power consumption.

Healthcare is being transformed by photonics, with advancements in imaging, diagnostics, and therapeutic applications. Optical coherence tomography (OCT) and multiphoton microscopy provide detailed insights into biological tissues, enabling early disease detection. Future applications could include wearable photonic devices for continuous health monitoring, advanced diagnostic tools for personalized medicine, and light-based therapies for targeted disease treatment.

Environmental Monitoring is benefiting from photonic sensors that offer real-time, high-sensitivity detection of air quality, water purity, and soil conditions. These sensors enable accurate monitoring of pollutants and other environmental factors. The future of environmental monitoring could involve networks of photonic sensors for comprehensive, continuous observation, supported by AI-driven data analysis and decision-making.

In manufacturing and industry, photonics is enabling high-precision processes such as laser machining, 3D printing, and optical metrology. These technologies offer speed and flexibility in the production of complex components and structures. Future applications could include fully automated production lines using photonic technologies, real-time quality control through advanced imaging systems, and the integration of photonics with robotics and the Internet of Things (IoT) for smart manufacturing.

Conclusion

The field of 3D photonics is advancing rapidly, driven by innovations in materials, fabrication techniques, and integration with emerging technologies. These developments are unlocking new possibilities across industries, from telecommunications and computing to healthcare and environmental monitoring. As researchers and engineers continue to push the boundaries of 3D photonics, the future promises a world where light-based technologies play an increasingly central role in enhancing efficiency, performance, and reliability across a wide range of applications. The integration of AI, flexible materials, and quantum photonics will further accelerate progress, shaping the next generation of photonic devices and systems.

10.2 Market Trends and Industry Outlook for 3D Photonics: Analysis of Current Market Dynamics and Future Growth Prospects

Introduction

The market for 3D photonics is expanding rapidly due to continuous technological advancements, growing demand across sectors such as telecommunications, data centers, healthcare, and consumer electronics, and the pressing need for higher performance, compactness, and efficiency in photonic devices. As industries push for faster data transmission, greater bandwidth, and reduced energy consumption, 3D photonics has become integral to addressing these needs through innovations in vertical integration, multi-layered photonic structures, and efficient light-guiding techniques. Key trends driving this market include the development of solid-state LiDAR for autonomous vehicles, advancements in quantum computing and sensing using 3D photonic architectures, and the integration of 3D photonic components with electronics for high-speed data processing. Despite its growth, the market faces challenges such as fabrication complexity, material compatibility issues, and high costs associated with advanced manufacturing. However, ongoing research and government support for photonic technologies, especially for data infrastructure and next-generation computing, present significant opportunities for market expansion. This section provides an in-depth look at these market dynamics, including key trends, drivers, and potential obstacles, while highlighting the future growth prospects as industries increasingly adopt 3D photonics to meet the demands of modern technology (Fig. 10.2).

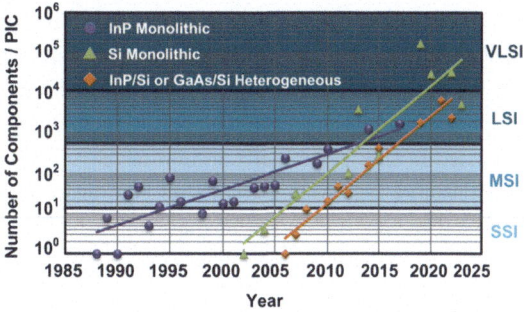

Fig. 10.2 Timeline showing emerging technologies and future applications enabled by 3D photonics. https://www.nature.com/articles/s41467-024-44750-0

Current Market Dynamics

The demand for high-speed communication is rapidly increasing, driven by the rise of 5G networks and the exponential growth of internet traffic. Telecommunication companies are investing heavily in photonic technologies to enhance bandwidth, reduce latency, and improve overall network performance. This growing need is accelerating the development and adoption of 3D photonic integrated circuits (PICs) and advanced optical fibers, contributing to a robust market expansion.

Advancements in manufacturing technologies are playing a crucial role in the evolution of 3D photonics. Innovations in nanofabrication, additive manufacturing, and lithography have enabled the production of complex 3D photonic structures with higher precision and efficiency. These breakthroughs are lowering production costs and improving scalability, making photonic devices more accessible across industries and further driving market growth.

The integration of photonics with emerging technologies such as artificial intelligence (AI), quantum computing, and the Internet of Things (IoT) is creating new opportunities for innovation. Photonic devices are becoming key enablers in these areas, offering high-speed data processing, low power consumption, and enhanced performance. This trend is expanding the application of photonics in smart cities, autonomous vehicles, and advanced medical diagnostics, fostering market expansion.

There has also been a significant increase in investments and funding for photonic research and development from both private and public sectors. Governments and organizations recognize the strategic importance of photonics in driving economic growth and technological advancement. Increased funding is accelerating research efforts, leading to new innovations and faster commercialization of 3D photonic technologies.

Market Drivers

The telecommunications and data center industries are major drivers of the 3D photonics market. The need for higher data transmission rates, improved network reliability, and lower energy consumption is pushing telecom companies toward photonic solutions. With the ongoing deployment of 5G networks and the expansion of cloud-based infrastructure, demand for photonic technologies is expected to grow steadily.

The healthcare sector is increasingly adopting photonic technologies for imaging, diagnostics, and therapeutic applications. Techniques such as optical coherence tomography (OCT) and multiphoton microscopy are becoming standard in medical diagnostics. As biophotonics continues to advance, the demand for 3D photonic devices in healthcare is expected to rise, particularly in non-invasive medical procedures.

Consumer electronics is another significant market segment benefiting from 3D photonics. The rapid adoption of augmented reality (AR) and virtual reality (VR) devices, along with high-resolution cameras and sensors, is driving the demand for advanced photonic components. As AR/VR applications continue to evolve, the consumer electronics sector is poised to contribute substantially to market growth.

10.2 Market Trends and Industry Outlook for 3D Photonics: Analysis ...

Environmental monitoring and industrial sensing applications are also expanding, with photonic sensors being used for real-time monitoring of air quality, water purity, and industrial processes. These sensors offer high sensitivity, accuracy, and real-time capabilities, making them ideal for ensuring environmental compliance. With increasing regulatory pressures and a growing emphasis on sustainability, the demand for photonic sensors is expected to continue rising.

Challenges

Despite the promising growth, the 3D photonics market faces several challenges. One major hurdle is the high initial cost associated with the development and deployment of photonic technologies. These expenses can act as a barrier to entry, particularly for small and medium-sized enterprises (SMEs), slowing adoption rates and limiting market penetration.

Manufacturing complexity is another challenge, as fabricating 3D photonic structures requires specialized equipment and expertise. Ensuring high precision and consistency in production can be difficult and time-consuming, leading to higher costs and extended development cycles.

Standardization and interoperability issues further complicate the adoption of 3D photonic devices. The lack of uniform standards for photonic components and systems makes integration across different industries and applications challenging. Addressing these concerns will be essential for enabling widespread adoption and ensuring seamless compatibility between various photonic technologies.

Future Growth Prospects

The expansion of 3D photonics in emerging markets presents significant opportunities. Regions such as Asia-Pacific, Latin America, and Africa are experiencing rapid industrialization and technological growth, driving the demand for advanced photonic solutions. Companies that establish a strong presence in these markets can tap into substantial growth potential and increase their market share.

Advancements in quantum photonics are expected to be a game-changer for multiple industries. Quantum photonic technologies are poised to revolutionize secure communication, computing, and ultra-sensitive sensing. Ongoing research and development in this field are likely to lead to groundbreaking innovations, opening new market segments and driving further expansion.

A growing focus on sustainability and energy efficiency is also fueling the adoption of photonic technologies. Photonics plays a critical role in renewable energy, green manufacturing, and sustainable agriculture. As industries strive to reduce their environmental footprint, photonic solutions that enable energy-efficient devices and processes will become increasingly valuable.

The development of collaborative innovation ecosystems, where academia, industry, and government entities work together, is accelerating the commercialization of 3D photonic technologies. These partnerships foster knowledge sharing, funding, and resource pooling, driving innovation and reducing time-to-market for new photonic solutions. By leveraging collaboration, companies can enhance their competitiveness and push the boundaries of photonic advancements. The market for 3D photonics is on a strong growth trajectory, driven by technological advancements, increasing demand across various sectors, and integration with emerging fields such as AI and quantum computing. While challenges such as high initial costs and manufacturing complexities remain, opportunities in emerging markets, quantum photonics, and sustainability-driven applications provide promising avenues for expansion. By addressing these challenges and capitalizing on key market drivers, the 3D photonics industry is well-positioned to shape the future of technology and innovation across multiple domains.

10.3 Potential Breakthroughs in 3D Photonics: Areas with the Highest Potential for Innovation and Development

Introduction

3D photonics is spearheading innovation across a range of industries by enabling new levels of data transmission speed, sensitivity, and miniaturization. In telecommunications, 3D photonics is enhancing data centers and network infrastructures with high-density photonic circuits that offer lower latency and higher bandwidth, critical for supporting 5G networks and beyond. In healthcare, 3D photonics has opened up possibilities for more compact and precise biomedical imaging devices, enabling early diagnosis and monitoring of diseases at a cellular level. Quantum computing is another field where 3D photonics is creating transformative opportunities, allowing for the development of highly integrated quantum circuits that leverage light for efficient and stable qubit manipulation. Meanwhile, in environmental monitoring, 3D photonic sensors with enhanced sensitivity are enabling real-time detection of pollutants and atmospheric changes, critical for climate studies and pollution control. This section examines the current advancements in these key areas, as well as the potential developments that could drive revolutionary changes in technology and society, from faster internet speeds and powerful computing to improved health diagnostics and environmental sustainability.

Quantum Photonics

Quantum computing leverages the principles of quantum mechanics to manipulate photons for information processing. Quantum computers use qubits that can exist in multiple states simultaneously, offering unprecedented computational power. Future breakthroughs

10.3 Potential Breakthroughs in 3D Photonics: Areas ...

could include scalable quantum photonic circuits that integrate numerous qubits, error-corrected quantum operations, and practical quantum algorithms for solving complex problems in cryptography, optimization, and materials science. These advancements could revolutionize computing by enabling solutions to problems that are currently intractable for classical computers, leading to progress in drug discovery, climate modeling, and artificial intelligence.

Quantum communication utilizes principles such as entanglement and superposition to create secure communication channels that are theoretically immune to eavesdropping. Key advancements could include long-distance quantum communication networks, quantum repeaters that extend the range of quantum signals, and integrated quantum photonic devices for robust and scalable quantum key distribution. These developments could lead to unhackable communication systems, enhancing security in critical applications such as financial transactions, governmental communications, and data centers.

Photonic Integrated Circuits (PICs)

Silicon photonics integrates optical components on a silicon chip, using the existing infrastructure of semiconductor manufacturing. This technology is already widely used in data centers and high-speed communication systems. Future developments could involve the integration of active components such as lasers and modulators on silicon, the development of low-loss waveguides, and advanced packaging techniques to improve performance and reduce costs. Enhanced silicon photonics could enable faster, more efficient data transmission, support the growth of cloud computing and IoT, and drive innovations in AI and machine learning by providing high-speed interconnects for data-intensive applications.

Hybrid photonic integration combines different materials and technologies on a single chip to exploit their unique optical properties, overcoming the limitations of silicon photonics. Future breakthroughs could involve the integration of materials like III-V semiconductors, lithium niobate, and graphene with silicon to create multifunctional photonic circuits with superior performance. This approach could lead to highly efficient and compact devices for telecommunications, sensing, and biomedical applications, paving the way for new technologies such as integrated LiDAR systems for autonomous vehicles.

Metamaterials and Metasurfaces

Advanced metamaterials are artificial structures engineered to exhibit properties not found in nature, such as negative refractive index and super-resolution imaging capabilities. They are used in applications such as lenses, cloaking devices, and antennas. Future developments could include tunable and reconfigurable metamaterials, dynamic control of optical properties, and scalable manufacturing techniques for large-area metamaterials. These advancements could enable next-generation imaging systems, highly efficient solar cells, and novel optical devices for communication and sensing, transforming industries ranging from healthcare to renewable energy.

Metasurfaces, the two-dimensional analogs of metamaterials, offer precise control over the phase, amplitude, and polarization of light. They are used in flat optics, holography, and beam shaping. Future innovations could lead to high-efficiency, multifunctional metasurfaces, integration with active components for dynamic control, and large-scale fabrication methods. These advancements could revolutionize optical systems by replacing bulky traditional optical elements with ultra-thin, planar devices, leading to compact and lightweight lenses, sensors, and display technologies.

Nonlinear Photonics

Nonlinear optical materials exploit the nonlinear interaction between light and matter to achieve effects such as frequency conversion, self-phase modulation, and soliton formation. Materials with strong nonlinear responses are critical for these applications. Future breakthroughs could involve the discovery and development of new nonlinear materials with enhanced properties, the engineering of nanostructures to boost nonlinear effects, and integration with photonic circuits. Enhanced nonlinear photonics could lead to the development of ultrafast optical switches, high-efficiency frequency converters for telecommunications, and advanced light sources for spectroscopy and imaging.

Nonlinear photonic devices are widely used in applications such as signal processing, wavelength conversion, and optical computing. These devices rely on the intrinsic nonlinear properties of the materials they are made from. Future innovations could include compact and efficient nonlinear devices, integration with silicon photonics platforms, and the development of tunable and programmable nonlinear photonic circuits. These advancements could enable all-optical signal processing, reducing reliance on electronic components and enhancing the speed and efficiency of communication networks and computing systems.

Biophotonics and Medical Applications

Advanced imaging techniques in biophotonics encompass a range of photonic technologies for imaging and diagnosing biological tissues. Techniques such as optical coherence tomography (OCT), multiphoton microscopy, and Raman spectroscopy are widely used. Future breakthroughs could include super-resolution imaging techniques, integration with AI for automated image analysis, and the development of portable and affordable imaging devices. Enhanced biophotonic imaging could revolutionize medical diagnostics by providing real-time, non-invasive, and high-resolution views of tissues, enabling early disease detection and personalized treatment plans.

Photonic therapeutic devices use light for treatment purposes, including photodynamic therapy, laser surgery, and light-activated drug delivery. These technologies provide targeted and minimally invasive treatment options. Future advancements could involve the development of new light-sensitive compounds, integration with smart sensors for real-time feedback, and wearable photonic devices for continuous therapy. These innovations

10.3 Potential Breakthroughs in 3D Photonics: Areas ...

could improve patient outcomes by offering precise and effective treatments with fewer side effects, particularly in oncology, dermatology, and neurology.

Environmental Monitoring and Sensing

Photonic sensors play a crucial role in environmental monitoring by detecting pollutants and measuring physical parameters such as temperature, pressure, and humidity. These sensors offer high sensitivity, specificity, and real-time monitoring capabilities. Future developments could involve the creation of multifunctional and miniaturized sensors, integration with wireless communication networks, and enhanced data analysis through AI. Advanced photonic sensors could address environmental challenges by enabling comprehensive and continuous monitoring of air, water, and soil quality, ensuring compliance with regulations and promoting sustainability.

Smart photonic networks integrate photonic sensors with data processing and communication technologies to create intelligent monitoring systems. These networks are used in smart cities, industrial automation, and environmental protection. Future innovations could lead to autonomous sensor networks, integration with IoT platforms, and the use of blockchain for secure data management. Smart photonic networks could enhance the efficiency and effectiveness of monitoring systems, providing real-time insights and enabling proactive measures to mitigate environmental risks and optimize resource management (Fig. 10.3).

Conclusion

The potential breakthroughs in 3D photonics span an impressive array of applications, promising to revolutionize fields like quantum computing, secure communication, advanced imaging, and environmental monitoring. In quantum computing, 3D photonic structures allow for the integration of densely packed qubits, potentially overcoming the limitations of traditional computing and enabling unprecedented processing power

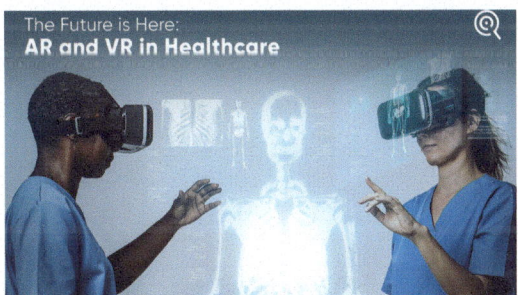

Fig. 10.3 Diagram of potential breakthroughs, highlighting high-impact areas such as augmented reality, healthcare, and quantum communications. https://www.linkedin.com/pulse/beyond-reality-exploring-therapeutic-potential-vr-ar-healthcare-cain/

for complex problem-solving. In quantum communication, 3D photonics can provide highly secure data transfer by leveraging quantum encryption methods, which are virtually immune to eavesdropping. For advanced imaging, 3D photonic technology supports compact, high-resolution imaging systems, such as those used in biomedical devices for early disease detection or in autonomous vehicles for precise object detection and navigation. In environmental monitoring, 3D photonic sensors enable real-time, highly sensitive detection of pollutants or greenhouse gases, offering powerful tools for climate research and pollution control. As research and development continue to advance, the field of 3D photonics is poised to address some of the world's most critical challenges, from data security and medical diagnostics to sustainable environmental practices, making it one of the most exciting and impactful areas of technological innovation today. The future holds a vast potential for these emerging applications, which could improve quality of life, drive economic growth, and create new technological frontiers.

Case Studies and Real-World Applications 11

- **Detailed Analysis of Successful Projects**: Examination of real-world implementations and their outcomes.
- **Lessons Learned and Best Practices**: Insights from industry leaders and researchers.
- **Potential Pitfalls and Solutions**: Common challenges and effective strategies to overcome them.

11.1 Detailed Analysis of Successful Projects in 3D Photonics: Examination of Real-World Implementations and Their Outcomes

Introduction

3D photonics has made substantial strides with real-world implementations that demonstrate its transformative potential across several high-impact industries, including telecommunications, healthcare, and environmental monitoring. In telecommunications, 3D photonic integration has enabled the development of compact, high-capacity optical transceivers and switches, which improve data transmission rates and reduce energy consumption, crucial for supporting the growing demands of data centers and 5G networks. In healthcare, 3D photonic devices have led to innovations in imaging systems, such as high-resolution endoscopes and optical coherence tomography (OCT) tools, which allow for early detection and precise monitoring of diseases. These advancements have improved diagnostic accuracy and patient outcomes by providing non-invasive, detailed imaging at the cellular level. Environmental monitoring has also benefited from 3D photonics

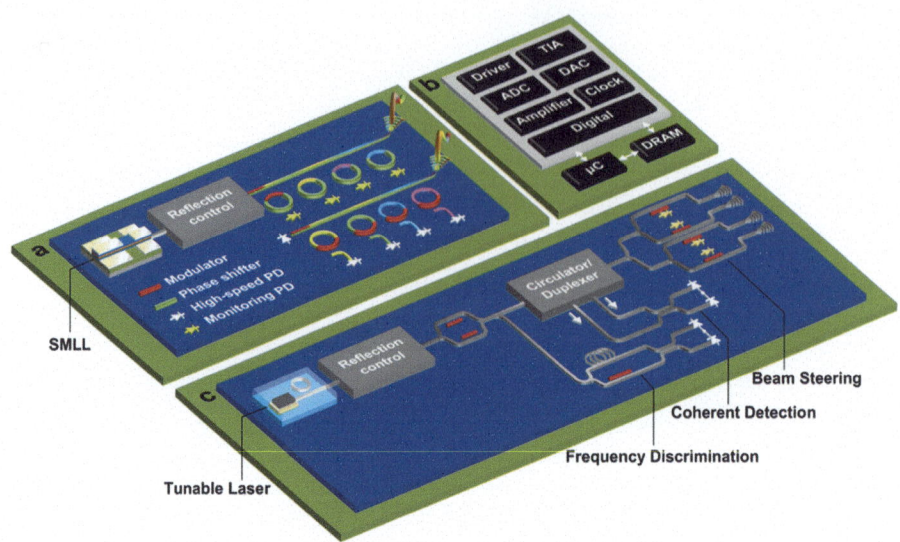

Fig. 11.1 Detailed diagram of a successful 3D photonic project, such as a LiDAR system or data center switch, showing components and performance metrics. https://www.nature.com/articles/s41467-024-44750-0

through the development of advanced sensors capable of detecting pollutants and greenhouse gases in real time, enhancing capabilities in climate research and pollution control. This section examines these groundbreaking projects, showcasing how 3D photonics is not only enhancing existing technologies but also paving the way for new applications, delivering tangible benefits and setting a foundation for future innovations in each of these fields (Fig. 11.1).

Project 1: Silicon Photonics for Data Centers

The objective of this project was to develop and deploy silicon photonic transceivers for high-speed data transmission in data centers, addressing the increasing demand for bandwidth and energy efficiency. The project involved the design and fabrication of silicon photonic integrated circuits (PICs) incorporating modulators, detectors, and waveguides. These transceivers were designed to support data rates of 100 Gbps and above, with a focus on scalability and cost-effectiveness.

The silicon photonic transceivers demonstrated significant improvements in data transmission speed and energy efficiency compared to traditional electronic transceivers. The integration of multiple optical components on a single chip reduced signal loss and enhanced overall performance. The use of silicon photonics enabled scalable production using existing semiconductor manufacturing processes, reducing costs and facilitating widespread adoption in data centers. As a result, companies such as Intel and Cisco have adopted silicon photonic technologies, driving further innovations in data communication.

Continued advancements in silicon photonics could lead to even higher data rates, lower power consumption, and more compact devices, further transforming the data center industry.

Project 2: Optical Coherence Tomography (OCT) in Medical Diagnostics

This project aimed to enhance medical imaging capabilities using Optical Coherence Tomography (OCT) for early detection and diagnosis of diseases such as glaucoma and macular degeneration. Researchers developed high-resolution OCT systems that use low-coherence light to capture cross-sectional images of biological tissues. The focus was on improving image resolution and acquisition speed to provide more detailed and real-time diagnostic information.

The enhanced OCT systems have been widely adopted in ophthalmology clinics, providing non-invasive, high-resolution imaging that allows for early detection of retinal diseases. The technology has also expanded into other medical fields, including dermatology and cardiology. Improved imaging capabilities have led to better diagnostic accuracy, enabling physicians to detect diseases at earlier stages and monitor disease progression more effectively. Early diagnosis and treatment have resulted in better patient outcomes, reducing the prevalence of severe complications and improving quality of life.

Future developments in OCT technology could further enhance its resolution and depth of imaging, making it applicable for a wider range of medical conditions and procedures.

Project 3: Photonic Sensors for Environmental Monitoring

This project focused on developing photonic sensors capable of real-time monitoring of environmental parameters such as air and water quality, with high sensitivity and specificity. Researchers created compact, portable sensors using photonic integrated circuits and advanced materials, designed to detect various pollutants and environmental factors with high accuracy.

The photonic sensors provided real-time data on environmental conditions, enabling rapid detection of pollutants and other hazardous substances. This capability has proven crucial in situations requiring immediate response, such as industrial spills or air quality monitoring during wildfires. The sensors demonstrated high sensitivity and specificity, ensuring accurate monitoring of even low concentrations of pollutants, which has facilitated better regulatory compliance and environmental protection measures. These sensors have been deployed in various settings, including urban environments, industrial sites, and natural reserves, contributing to comprehensive environmental monitoring networks.

Future projects could integrate these photonic sensors with Internet of Things (IoT) platforms to create smart environmental monitoring systems, providing real-time data and analytics for better environmental management.

Project 4: Metamaterials for Advanced Imaging Systems

The goal of this project was to develop metamaterials-based imaging systems that surpass the diffraction limit, enabling super-resolution imaging for applications in microscopy and security. Researchers designed and fabricated metamaterials with tailored optical properties, such as negative refractive index, using advanced nanofabrication techniques to achieve high precision.

The metamaterials-based imaging systems achieved resolutions beyond the diffraction limit, providing unprecedented detail in microscopic imaging. This advancement has significant implications for biological research, allowing scientists to observe cellular and molecular structures with greater clarity. The enhanced imaging capabilities have also been applied in security and surveillance, providing clearer and more detailed images for threat detection and analysis. Companies and research institutions have begun commercializing these advanced imaging systems, making the technology more accessible for various scientific and industrial applications.

Continued improvements in metamaterials and imaging technologies could lead to broader adoption in fields such as healthcare, materials science, and forensic analysis, driving further innovations and applications.

Project 5: 3D Printed Photonic Devices

This project explored the use of additive manufacturing techniques to create custom 3D photonic devices with complex geometries and tailored optical properties. Researchers utilized high-resolution 3D printing technologies to fabricate photonic devices such as waveguides, sensors, and micro-optical components, focusing on optimizing the printing process and material selection to achieve high performance.

The ability to 3D print photonic devices allowed for the customization of optical components to meet specific application requirements. This capability has been particularly beneficial in creating bespoke solutions for niche applications in research and industry. The use of 3D printing enabled rapid prototyping and iteration, reducing development time and costs, thereby accelerating the innovation cycle in photonics and allowing for faster commercialization of new technologies. Applications of 3D printed photonic devices span telecommunications, medical diagnostics, and environmental sensing, demonstrating their versatility and effectiveness.

Future advancements in 3D printing materials and techniques could further enhance the capabilities of printed photonic devices, enabling more complex and high-performance optical systems.

Conclusion

These successful projects in 3D photonics highlight the transformative potential of photonic technologies across various industries. By addressing critical challenges and leveraging advanced methodologies, these projects have demonstrated significant improvements in performance, efficiency, and application scope. The outcomes of these real-world implementations not only underscore the current capabilities of 3D photonics but also pave the way for future innovations and broader adoption. As research and development in this field continue to advance, the impact of 3D photonics is poised to grow, driving further technological breakthroughs and enhancing our ability to address complex challenges in communication, healthcare, environmental monitoring, and beyond.

11.2 Lessons Learned and Best Practices: Insights from Industry Leaders and Researchers Working in Photonics

Introduction

The field of 3D photonics is marked by rapid advancements, driven by interdisciplinary collaborations, cutting-edge research, and a wide range of practical applications in sectors like telecommunications, healthcare, and environmental monitoring. Insights from industry leaders and researchers play a crucial role in guiding these developments, offering lessons on innovation, collaboration, and best practices that can maximize the impact of photonic technologies. In particular, innovation within 3D photonics is encouraged by continuously exploring new materials, advanced fabrication techniques, and complex integration processes, which push the boundaries of performance and functionality. Collaboration across disciplines—such as physics, engineering, materials science, and data science—has proven essential for tackling complex challenges in photonic integration, enabling teams to develop scalable, high-performance devices. Commercialization insights emphasize the importance of aligning technological developments with market needs, ensuring that 3D photonic solutions are not only advanced but also accessible and cost-effective. Additionally, the focus on sustainability encourages researchers to design energy-efficient devices, use eco-friendly materials, and create solutions that reduce environmental impact, which is increasingly critical as photonic technologies become more integral to modern infrastructure. This section explores these key insights and best practices, providing a roadmap for how future research and development efforts can continue to advance 3D photonics and expand its transformative potential across various industries (Fig. 11.2).

Fostering Innovation

Innovation in photonics often occurs at the intersection of multiple disciplines, including physics, materials science, engineering, and computer science. Industry leaders emphasize the importance of fostering interdisciplinary collaboration to drive breakthroughs. Encouraging cross-disciplinary teams and projects that combine expertise from various fields can facilitate creative problem-solving and innovation. Promoting a culture of open communication and knowledge sharing further enhances the potential for groundbreaking discoveries.

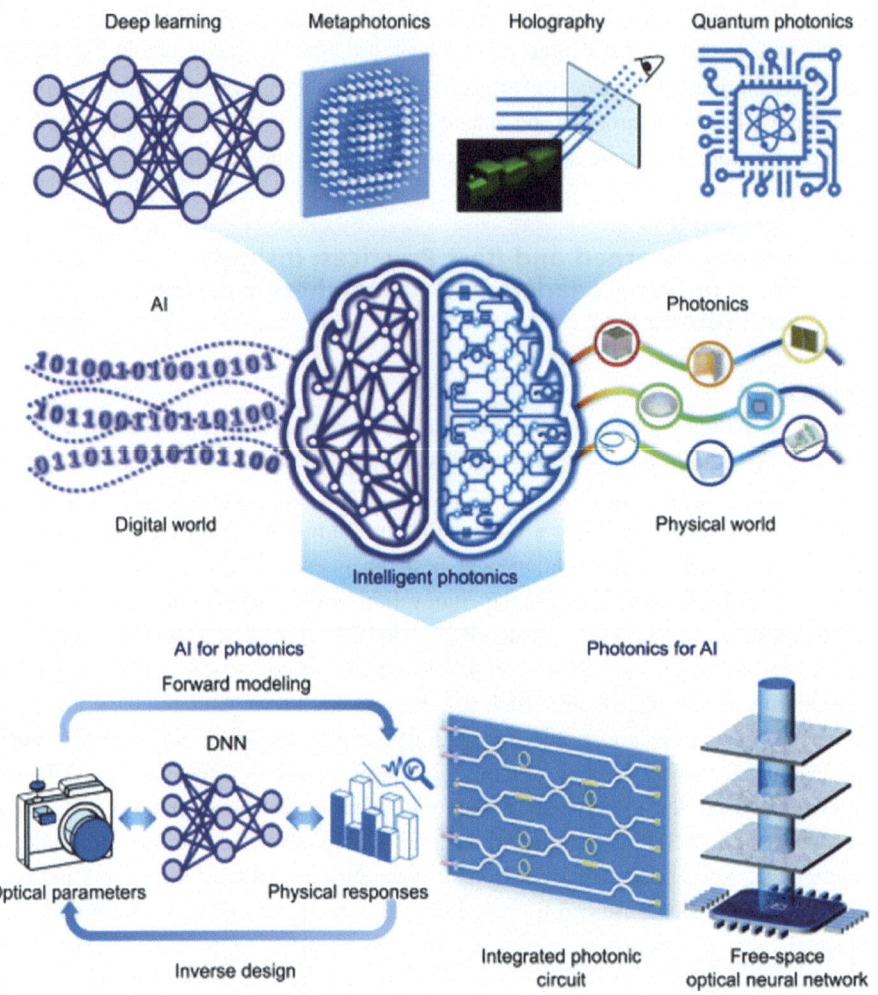

Fig. 11.2 Infographic of lessons learned and best practices from real-world implementations. https://www.sciencedirect.com/science/article/pii/S2095809924005149

Continuous investment in research and development (R&D) is crucial for maintaining a competitive edge in photonics. Companies and institutions that prioritize R&D are better positioned to develop cutting-edge technologies and respond to market demands. Allocating significant resources to R&D initiatives and supporting long-term projects that may not yield immediate returns but have high potential for transformative impact can ensure sustained growth. Establishing partnerships with academic institutions allows organizations to leverage their research capabilities and stay ahead of emerging trends.

Advanced simulation and modeling tools are essential for predicting the behavior of photonic structures and optimizing designs. These tools can significantly reduce development time and costs by identifying potential issues early in the design process. Incorporating state-of-the-art simulation and modeling software into the development workflow and training engineers and researchers to effectively use these tools ensures more accurate design decisions and streamlined innovation.

Enhancing Collaboration

Collaboration between industry and academia is a powerful driver of innovation in photonics. Academic institutions provide foundational research and access to emerging talent, while industry partners offer practical insights and resources for commercialization. Establishing formal partnerships with universities and research institutions supports joint research projects, internships, and collaborative workshops, fostering a continuous exchange of ideas and expertise.

Open innovation, where organizations collaborate with external partners to co-create solutions, can accelerate the development of new technologies. Participating in industry consortia, research networks, and innovation hubs promotes open collaboration, enabling companies to tap into a wider pool of knowledge and resources. Encouraging employees to engage with external experts, attend conferences, and contribute to collaborative research initiatives fosters a culture of continuous learning and growth.

Knowledge sharing within and across organizations enhances collective learning and innovation. Industry leaders highlight the importance of creating environments where knowledge flows freely and employees are encouraged to share insights and experiences. Implementing knowledge management systems that capture and disseminate critical information ensures that valuable expertise is accessible to all stakeholders. Organizing regular seminars, workshops, and cross-functional meetings facilitates the exchange of ideas and best practices, driving further advancements in photonics.

Successful Commercialization

Successful commercialization of photonic technologies requires a deep understanding of market needs and customer pain points. Companies that align their R&D efforts with market demands are more likely to achieve commercial success. Conducting thorough market research to identify emerging trends, customer requirements, and competitive landscapes provides valuable insights for product development. Engaging with potential customers early in the development process allows for feedback collection and validation of product concepts, ensuring a better market fit.

In the fast-paced photonics industry, the ability to bring products to market quickly is a competitive advantage. Companies that streamline their development processes and reduce time-to-market can capture market opportunities more effectively. Implementing agile development methodologies that emphasize iterative design, rapid prototyping, and frequent testing enhances flexibility and responsiveness to changing market conditions.

Photonic devices must meet stringent quality and reliability standards, especially in critical applications such as healthcare and telecommunications. Ensuring high product quality is essential for building customer trust and achieving long-term success. Establishing rigorous quality control processes throughout the development and manufacturing stages minimizes defects and performance variability. Investing in advanced testing and validation equipment further ensures that products meet industry standards and customer expectations.

Prioritizing Sustainability

Sustainability is becoming increasingly important in the photonics industry. Companies that adopt environmentally friendly manufacturing practices can reduce their environmental footprint and appeal to environmentally conscious customers. Implementing sustainable manufacturing processes that minimize waste, reduce energy consumption, and use eco-friendly materials contributes to both environmental and economic benefits. Investing in technologies that enable the recycling and reuse of photonic components enhances long-term sustainability efforts.

Energy efficiency is a critical consideration for many photonic applications, particularly in data centers and telecommunications. Developing energy-efficient photonic devices can lead to significant cost savings and environmental benefits. Focusing R&D efforts on creating photonic devices with lower power consumption and higher efficiency ensures alignment with industry demands. Collaborating with customers to understand their energy efficiency requirements allows companies to design products that meet or exceed these standards.

The circular economy model, which emphasizes the reuse, repair, and recycling of products, is gaining traction in the photonics industry. Companies that embrace these principles can reduce waste and create sustainable business models. Designing photonic devices with modular components that can be easily repaired or upgraded extends product lifecycles and reduces electronic waste. Developing recycling programs that enable the recovery and reuse of valuable materials from end-of-life products further strengthens sustainability initiatives.

Case Studies of Best Practices

Intel has been a pioneer in silicon photonics, focusing on integrating optical and electronic components on a single chip to enhance data center performance. The company's success is attributed to its strong R&D investment, strategic partnerships with academia, and commitment to understanding market needs. Intel has also emphasized sustainable manufacturing and energy-efficient design, leading to widespread adoption of its silicon photonics products in data centers, enabling faster and more energy-efficient data transmission.

In the medical field, a collaborative project between a leading medical device company and a university developed an advanced Optical Coherence Tomography (OCT) system for early cancer detection. The project leveraged interdisciplinary collaboration, combining expertise in photonics, biomedical engineering, and clinical research. Rigorous testing and validation ensured high reliability and accuracy, resulting in a successfully commercialized OCT system that provides a non-invasive and highly effective tool for early cancer diagnosis, leading to better patient outcomes.

Conclusion

The lessons learned and best practices from industry leaders and researchers in the field of 3D photonics highlight the importance of innovation, collaboration, commercialization, and sustainability. By embracing interdisciplinary approaches, fostering strong partnerships, and prioritizing quality and environmental responsibility, companies and institutions can drive significant advancements in photonic technologies. These insights provide a roadmap for future success, enabling the continued growth and impact of 3D photonics across various industries.

11.3 Potential Pitfalls and Solutions: 3D Photonics Common Challenges and Effective Strategies to Overcome Them

Introduction
The field of 3D photonics holds tremendous potential for transformative advancements across industries such as telecommunications, computing, and healthcare. However, the development of 3D photonic technologies faces several challenges that could slow progress if not addressed effectively. One major challenge is the complexity of fabrication, as creating multi-layered photonic structures requires precise alignment, advanced materials, and sophisticated integration techniques. Misalignment or defects at even the smallest scale can disrupt optical performance, making quality control and precision manufacturing critical. Thermal management is another significant hurdle; densely packed 3D photonic circuits generate heat, which can affect performance and longevity, requiring advanced cooling methods and materials with high thermal conductivity. Additionally, material compatibility remains an obstacle, as photonic and electronic components often require different substrates or fabrication conditions, complicating the integration process in hybrid systems.

To address these challenges, researchers and industry leaders are exploring innovative solutions, such as automated alignment tools, thermal interface materials, and co-design methodologies that consider both photonic and electronic requirements from the outset. Furthermore, collaboration across disciplines—including materials science, engineering, and computational modeling—enables more comprehensive approaches to these issues. By identifying and addressing these potential pitfalls, this section highlights strategies that can support the continued growth and success of 3D photonics, ensuring that it reaches its full potential to impact next-generation technologies (Fig. 11.3).

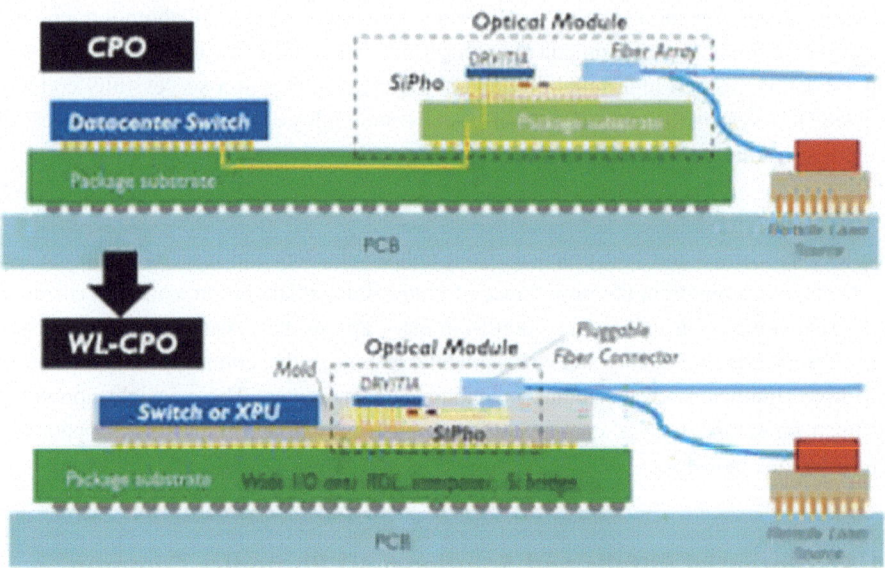

Fig. 11.3 Flowchart of potential pitfalls in 3D photonic development, along with recommended solutions. https://www.electronicdesign.com/technologies/embedded/article/21270649/electronic-design-silicon-photonics-will-shine-in-the-age-of-ai

Technical Challenges

The fabrication of 3D photonic structures involves highly complex processes that require precise control over material properties and geometries. This complexity can lead to high production costs and long development times. Investing in advanced fabrication techniques such as electron-beam lithography, two-photon polymerization, and direct laser writing can offer greater precision and scalability. Additionally, adopting modular design principles can simplify fabrication by allowing for the assembly of smaller, more manageable components into larger systems.

The performance of 3D photonic devices is heavily dependent on the properties of the materials used. Current materials may have limitations in terms of refractive index, absorption, and thermal stability, which can constrain device performance. Engaging in continuous research and development to discover and optimize new materials with superior properties is essential. Collaborating with materials scientists to explore novel compounds and composites, as well as utilizing hybrid integration of multiple materials, can combine their strengths and enhance overall device performance.

Photonic devices can generate significant heat during operation, leading to performance degradation and reliability issues. Effective thermal management is essential to maintain device functionality. Designing devices with integrated heat sinks and thermal interface materials can help dissipate heat efficiently. Utilizing materials with high thermal conductivity and exploring passive cooling techniques such as microfluidic channels can prevent overheating and improve reliability.

Manufacturing and Scalability
The cost of producing 3D photonic devices can be prohibitively high, especially for small-scale production, which can limit their adoption in cost-sensitive markets. Scaling production to benefit from economies of scale can help reduce per-unit costs. Automating manufacturing processes increases efficiency and reduces labor costs, while partnering with established semiconductor foundries leverages existing infrastructure and expertise, further lowering production costs.

Ensuring high yield and consistent quality in the production of 3D photonic devices is challenging due to the precision required in fabrication. Variability in manufacturing processes can lead to defects and performance inconsistencies. Implementing rigorous quality control protocols throughout the manufacturing process, utilizing advanced inspection and metrology tools to detect and correct defects early, and adopting statistical process control (SPC) techniques can help monitor and maintain production quality, ensuring high yield and consistent device performance.

Scaling up from prototype to mass production poses significant challenges in maintaining performance and reliability while controlling costs. Developing scalable manufacturing techniques that can be easily adapted from lab-scale to industrial-scale production is crucial. Collaborating with industry partners to share knowledge and resources facilitates the transition to mass production. Investing in pilot production lines allows companies to identify and address scalability issues before full-scale manufacturing.

Integration and Interoperability
Integrating 3D photonic devices with existing electronic and optical systems can be complex and requires careful consideration of compatibility and performance. Designing photonic devices with standardized interfaces and protocols ensures compatibility with existing systems. Collaborating with system integrators during the design phase can help address potential integration issues early. Utilizing co-packaging techniques to combine photonic and electronic components within a single package enhances integration and performance.

A lack of standardization and interoperability between different photonic components and systems can hinder widespread adoption and integration. Advocating for the development and adoption of industry-wide standards for photonic components and systems is essential. Participating in standardization bodies and consortia can help influence the creation of interoperability standards. Designing devices with flexibility and adaptability in mind also facilitates integration with diverse systems, ensuring smoother adoption across various industries.

Market and Adoption

The market for 3D photonic devices is still emerging, with uncertain demand and evolving applications. This uncertainty can make it challenging to justify investment in research and development. Conducting thorough market research to identify emerging trends, potential applications, and customer needs is crucial. Engaging with early adopters and industry stakeholders to validate market potential and gather feedback can help mitigate risks. Developing flexible and versatile products that serve multiple applications can also reduce market uncertainty and broaden adoption.

Potential customers may lack awareness or understanding of the benefits and capabilities of 3D photonic devices, hindering adoption. Investing in customer education and outreach programs can demonstrate the value and potential of photonic technologies. Providing comprehensive training and support helps customers integrate photonic devices into their applications. Case studies, demonstrations, and pilot projects can effectively showcase the benefits, increasing confidence and accelerating market adoption.

Navigating regulatory and compliance requirements can be challenging, particularly in highly regulated industries such as healthcare and telecommunications. Engaging with regulatory bodies early in the development process ensures that requirements are understood and compliance is met. Investing in regulatory expertise can streamline the approval process. Building devices with robust safety and performance features facilitates regulatory approval and market entry, ensuring faster commercialization and industry acceptance.

Research and Development

The field of photonics is rapidly evolving, with new technologies and discoveries emerging at a fast pace. Keeping up with these changes can be challenging for researchers and companies. Fostering a culture of continuous learning and innovation within organizations is essential. Encouraging researchers to stay abreast of the latest developments through conferences, journals, and collaborations enhances knowledge sharing. Establishing flexible R&D processes that quickly adapt to new technologies ensures that organizations remain competitive and at the forefront of innovation.

Securing adequate funding and resources for long-term research projects can be difficult, especially in the face of competing priorities and budget constraints. Diversifying funding sources by pursuing grants, partnerships, and collaborations with industry and government agencies can provide financial stability. Prioritizing projects with the highest potential impact and aligning them with strategic goals ensures that resources are allocated effectively. Efficient project management can maximize the return on investment for R&D activities, ensuring sustainable innovation.

Attracting and retaining top talent in the competitive field of photonics can be challenging, particularly for startups and smaller companies. Offering competitive compensation packages and creating a stimulating work environment that encourages innovation and professional growth can attract skilled professionals. Providing opportunities for continuous learning and career advancement enhances job satisfaction and retention. Building a strong organizational culture that values and recognizes contributions fosters loyalty, ensuring that top talent remains engaged and committed to driving technological advancements.

Conclusion

The field of 3D photonics offers remarkable potential for driving innovations across industries, promising breakthroughs in data communications, sensing, healthcare, and more. However, achieving this potential requires overcoming significant challenges, such as complex fabrication processes, thermal management, and material compatibility issues that can hinder the performance and scalability of 3D photonic devices. Addressing these challenges effectively demands a multifaceted approach that brings together expertise from diverse fields, including materials science, engineering, and computational modeling. Embracing interdisciplinary collaboration allows researchers and engineers to pool knowledge and resources, accelerating problem-solving and the development of robust solutions. Furthermore, fostering a culture of quality and sustainability is essential to ensure the long-term reliability and environmental viability of these advanced photonic systems. This includes investing in quality control, green manufacturing techniques, and materials that minimize environmental impact. Lastly, staying responsive to market needs is critical; by aligning research goals with industry demands, 3D photonics can better address practical applications, ensuring the technology's relevance and commercial success. By navigating these complexities with strategic foresight, 3D photonics has the potential to achieve transformative, long-term impact across multiple high-tech fields.

Vision for the Future 12

- **Summary of Key Points**: Recap of the main topics covered in the book.
- **Future Impact of 3D Photonics**: Predictions for how 3D photonics will shape future technologies.
- **Final Thoughts and Forward-Looking Statements**: Author's perspective on the evolving landscape of 3D photonics.

12.1 Summary of Key Points in 3D Photonics: Recap of the Main Topics Covered in the Book

Introduction
3D photonics marks a transformative step in technology, paving the way for innovations across fields as diverse as telecommunications, computing, healthcare, and environmental monitoring. This book has provided an in-depth exploration of the multifaceted landscape of 3D photonics, beginning with its technical foundations, which include advanced fabrication techniques, integration strategies, and challenges in managing optical and thermal properties in multi-layered structures. The discussion then moved to market dynamics, examining the growing demand for compact, high-performance photonic devices and the industry trends shaping future applications. Through case studies and examples of successful implementations, readers gained insight into real-world uses of 3D photonics in data centers, quantum computing, biomedical imaging, and environmental sensors. Finally, the book considered the future prospects of this rapidly evolving field, highlighting emerging innovations in materials, design methodologies, and potential applications that could redefine entire industries. This section recaps the book's essential insights

Fig. 12.1 Visionary illustration of future 3D photonic technologies, such as integrated photonic chips in various futuristic devices. https://insidetelecom.com/photonic-integrated-circuits-the-future-of-computing/

and conclusions, offering a comprehensive overview of how 3D photonics is set to revolutionize technology and improve the quality of life in the years to come (Fig. 12.1).

Foundations of 3D Photonics

The book begins by introducing the fundamental principles of photonics, including the behavior of light, waveguide technology, and the role of materials with varying refractive indices. Key technologies such as photonic crystals, metamaterials, and integrated photonic circuits are explained, providing a strong foundation for understanding subsequent advancements. This foundational knowledge is crucial for grasping how photonic devices manipulate light to achieve desired functionalities.

A detailed exploration of fabrication techniques follows, covering advanced methods such as electron-beam lithography, two-photon polymerization, and direct laser writing. These techniques enable the creation of intricate 3D photonic structures with high precision. The ability to fabricate complex photonic devices is fundamental to the implementation of 3D photonics, as advancements in fabrication significantly impact scalability, performance, and cost-effectiveness.

Market Trends and Industry Dynamics

The book analyzes the current market dynamics, revealing a growing demand for photonic technologies in telecommunications, data centers, healthcare, and consumer electronics. The integration of photonics with emerging technologies like AI and IoT is also

12.1 Summary of Key Points in 3D Photonics: Recap of the Main ...

highlighted. Recognizing market trends helps stakeholders understand the driving forces behind the adoption of 3D photonics and identify areas with the highest growth potential. Market dynamics influence strategic decisions regarding research, development, and commercialization.

The industry outlook discusses future growth prospects, including expansion into emerging markets, advancements in quantum photonics, and an increasing focus on sustainability. Predictions on how these factors will shape the industry are provided, helping businesses and researchers align their efforts with future opportunities to remain competitive in a rapidly evolving field.

Technical Advancements and Innovations

Cutting-edge developments in 3D photonics are explored, such as quantum photonics, hybrid photonic integration, and advanced metamaterials. The potential applications and transformative impact of these technologies are discussed, emphasizing the importance of staying informed about emerging innovations. These advancements represent the frontier of what is possible with 3D photonics, offering new solutions to complex challenges and opening up new application areas.

Predictive modeling is another key area covered, with techniques such as finite-difference time-domain (FDTD) and finite element method (FEM) examined. These methods are essential for optimizing the performance and reliability of photonic devices. Predictive modeling helps design efficient and effective photonic structures, reducing development time and costs while enabling engineers to anticipate and mitigate potential issues before they arise.

Successful Implementations

Several real-world projects showcasing successful implementations of 3D photonics in various sectors are analyzed. These case studies highlight the objectives, methodologies, outcomes, and impact of each project, providing valuable insights into best practices, challenges, and strategies for success. Learning from these implementations offers benchmarks for future projects and demonstrates the tangible benefits of 3D photonics.

The impact of 3D photonics across industries is examined in detail, particularly its role in telecommunications through silicon photonics, healthcare via optical coherence tomography, and environmental monitoring with photonic sensors. Understanding the influence of 3D photonics on various industries underscores its versatility and potential, showcasing its broad applicability and ability to address diverse needs and challenges.

Challenges and Solutions

Several challenges in the field of 3D photonics are identified, including complex fabrication processes, material limitations, thermal management, and high production costs. These obstacles can hinder the development and adoption of photonic technologies. Recognizing these challenges is the first step toward addressing them, enabling researchers and industry leaders to develop targeted strategies to overcome barriers and ensure continued progress.

Various solutions to these challenges are proposed, such as investing in advanced fabrication techniques, developing new materials, implementing rigorous quality control, and fostering interdisciplinary collaboration. Effective strategies are crucial for mitigating challenges and unlocking the full potential of 3D photonics. These strategies provide a roadmap for researchers and companies, facilitating both innovation and commercialization.

Best Practices and Insights

Industry leaders and researchers share insights emphasizing the importance of interdisciplinary approaches, continuous R&D investment, and strong industry-academia partnerships. Learning from these experiences provides practical guidance for navigating the complexities of 3D photonics and achieving success. These insights offer valuable lessons that can be applied across various projects and research initiatives.

The book also discusses strategies for successful commercialization, including understanding market needs, accelerating time-to-market, and ensuring product quality and reliability. Effective commercialization strategies are essential for bringing photonic technologies from the lab to the market. These strategies help ensure that innovative products meet customer needs and achieve commercial success.

Future Prospects and Innovations

Quantum photonics is explored as a major frontier in technology, with the potential to revolutionize computing and communication. Advances in quantum photonic circuits and quantum communication networks are highlighted, showing how quantum photonics could unlock new levels of performance and security. Understanding these prospects helps guide future research and investment, positioning organizations to lead in this emerging field.

The importance of developing sustainable photonic technologies is also emphasized, focusing on energy-efficient designs, sustainable manufacturing practices, and circular economy principles. Sustainability is becoming increasingly important in technology development, and integrating sustainable practices ensures that photonic technologies contribute positively to environmental and societal goals.

Conclusion

This book provides a comprehensive overview of 3D photonics, covering its foundational principles, market dynamics, technical advancements, successful implementations, challenges, and future prospects. The insights and strategies discussed offer valuable guidance for researchers, engineers, and industry leaders navigating the complexities of this rapidly evolving field. As 3D photonics continues to advance, it promises to drive significant technological innovations and address some of the most pressing challenges across various industries. The knowledge and best practices shared aim to support and inspire ongoing efforts in the exciting journey of 3D photonics.

12.2 Future Impact of 3D Photonics: Predictions for How 3D Photonics Will Shape Future Technologies

Introduction

3D photonics, with its advanced ability to manipulate and control light at microscopic and nanoscopic scales, is on the brink of revolutionizing multiple industries by enabling faster, more efficient, and highly integrated photonic systems. As research progresses and fabrication techniques become more refined, 3D photonics is set to drive breakthroughs in telecommunications, where ultra-compact, high-speed data transmission systems can support the growing demands of 5G and 6G networks. In computing, 3D photonics can enable ultra-fast data processing and low-energy optical interconnects, which are crucial for developing high-performance data centers and AI systems. In healthcare, 3D photonic devices are expected to enhance medical imaging, diagnostics, and surgical tools by providing high-resolution, non-invasive imaging techniques at the cellular level. Furthermore, in environmental monitoring, the sensitivity of 3D photonic sensors can be applied to real-time detection of pollutants and greenhouse gases, supporting climate research and pollution control initiatives. This section discusses these potential advancements, forecasting how 3D photonics will shape future technologies and address some of the world's most pressing challenges by enabling more precise, compact, and energy-efficient solutions across diverse applications.

Telecommunications and Data Transmission

3D photonics is set to revolutionize telecommunications by enabling ultra-high-speed data transmission. Photonic integrated circuits (PICs) and silicon photonics will support data rates exceeding terabits per second, far surpassing the capabilities of current electronic systems. This advancement will lead to faster internet speeds, more efficient data centers, and improved global connectivity, facilitating the growth of data intensive applications such as virtual reality, augmented reality, and 8K streaming.

The deployment of 5G networks will be significantly enhanced by 3D photonics, providing the backbone for even more advanced communication technologies and potentially leading to the development of 6G networks. Enhanced network performance will enable real-time communication and data transfer for autonomous vehicles, smart cities, and the Internet of Things (IoT). The increased bandwidth and reduced latency will support innovative applications in various fields, from industrial automation to telemedicine.

3D photonics will also be integral to the development of secure quantum communication networks. Quantum key distribution (QKD) systems, leveraging photonic technologies, will ensure unbreakable encryption. These advancements will enhance data security for financial transactions, governmental communications, and critical infrastructure, leading to widespread adoption of quantum-safe encryption methods and safeguarding sensitive information against future threats posed by quantum computing.

Computing and Information Processing

Photonic computing is emerging as a viable alternative to traditional electronic computing, offering significantly higher processing speeds and lower power consumption. This shift will revolutionize fields requiring massive computational power, such as artificial intelligence (AI), machine learning, and big data analytics. Photonic processors will enable more complex AI models and faster data processing, driving advancements in autonomous systems, natural language processing, and predictive analytics.

3D photonics will play a crucial role in the development of scalable quantum computers. Photonic qubits and integrated quantum photonic circuits will form the basis of next-generation quantum processors, unlocking new capabilities in cryptography, materials science, drug discovery, and climate modeling. The ability to perform computations that are currently infeasible for classical computers will lead to breakthroughs in solving complex scientific and engineering problems.

Optical interconnects, enabled by 3D photonics, will replace electronic interconnects in data centers and high-performance computing systems, facilitating faster and more efficient data transfer. This transition will reduce energy consumption and heat generation associated with electronic interconnects, improving the sustainability and performance of computing systems. Enhanced data transfer rates will support the growing demands of cloud computing and data-intensive applications.

Healthcare and Biomedical Applications

3D photonics will drive advancements in medical imaging technologies, providing higher resolution, faster imaging, and real-time diagnostic capabilities. Improved imaging techniques, such as optical coherence tomography (OCT) and multiphoton microscopy, will enable early detection and precise monitoring of diseases. These advancements will lead to better patient outcomes, more effective treatments, and personalized medicine.

Non-invasive diagnostic tools leveraging 3D photonics will become more prevalent, allowing for real-time monitoring of physiological parameters and early detection of health issues. These tools will revolutionize healthcare by providing patients with more comfortable and convenient diagnostic options. Continuous health monitoring and early intervention will reduce healthcare costs and improve overall health outcomes.

Photonic therapeutic devices, such as laser surgery tools and photodynamic therapy systems, will become more advanced and widely adopted. These devices will offer minimally invasive treatment options with high precision, reducing recovery times and improving patient safety. Photonic therapies will be used to treat a wide range of conditions, from cancer to skin disorders, enhancing the effectiveness of medical treatments.

Environmental Monitoring and Sustainability

3D photonic sensors will enable real-time monitoring of environmental parameters, such as air and water quality, with unprecedented sensitivity and accuracy. This capability will facilitate better environmental management and regulatory compliance, enabling rapid response to environmental hazards and improving public health and safety. Enhanced monitoring will also support efforts to combat climate change and protect natural resources.

The development of energy-efficient photonic devices will contribute to more sustainable technologies and reduce the carbon footprint of various industries. Energy-efficient lighting, displays, and communication systems will lower energy consumption and greenhouse gas emissions. Photonic technologies will play a key role in creating sustainable infrastructure and promoting environmental stewardship.

3D photonics will be integrated into smart agriculture systems, providing precise monitoring and control of agricultural processes. Enhanced sensing and imaging technologies will optimize irrigation, fertilization, and pest management, increasing crop yields and reducing resource waste. Smart agriculture will contribute to food security and sustainable farming practices.

Advanced Manufacturing and Industrial Applications

3D photonic technologies will revolutionize precision manufacturing processes, enabling the production of complex structures with high accuracy and efficiency. Industries such as aerospace, automotive, and electronics will benefit from improved manufacturing capabilities. Photonic techniques will reduce material waste, lower production costs, and enhance product quality.

3D printing with photonic materials will become more advanced, allowing for the creation of custom photonic devices and components. Additive manufacturing will enable rapid prototyping and on-demand production of photonic devices, fostering innovation and reducing time-to-market. Customization will meet specific application needs, driving advancements in various fields.

Photonic sensors and automation systems will enhance industrial processes, providing real-time monitoring and control. Improved sensing capabilities will optimize production efficiency, reduce downtime, and ensure product quality. Automation systems will enhance operational safety and reduce labor costs, contributing to the competitiveness of industrial sectors.

The future impact of 3D photonics is vast and multifaceted, with the potential to transform numerous industries and improve quality of life. Predictions for the field indicate significant advancements in telecommunications, computing, healthcare, environmental monitoring, and manufacturing. By enabling ultra-high-speed data transmission, advanced medical diagnostics, real-time environmental monitoring, and precision manufacturing, 3D photonics will drive technological progress and address critical global challenges.

As researchers, engineers, and industry leaders continue to explore and develop photonic technologies, the transformative potential of 3D photonics will become increasingly evident, shaping the future of technology and innovation.

12.3 Forward-Looking Statements: Author's Perspective on the Evolving Landscape of 3D Photonics

Introduction

As we conclude our exploration into the expansive field of 3D photonics, it's evident that this technology is on the verge of revolutionizing numerous industries with its unique ability to integrate complex photonic components into compact, multi-layered structures. This journey has highlighted not only the technical principles and breakthroughs that underpin 3D photonics but also the market dynamics that drive its growing adoption across fields like telecommunications, computing, healthcare, and environmental monitoring. Each chapter has illustrated how advancements in 3D photonic fabrication, integration, and thermal management techniques have enabled the creation of highly efficient, scalable devices capable of meeting modern demands for speed, miniaturization, and energy efficiency.

The final thoughts in this section reflect on the significant progress made in recent years, while also casting an eye toward the future, where continued innovation and interdisciplinary collaboration will be critical in overcoming remaining challenges, such as fabrication complexity and alignment precision. As 3D photonics continues to evolve, it holds the potential to transform entire sectors, paving the way for next-generation technologies in data transmission, high-performance computing, medical diagnostics, and environmental sensing. The possibilities are vast, and as research and development efforts accelerate, 3D photonics is poised to not only enhance current technologies but also create entirely new applications, solidifying its role as a catalyst for technological progress in the years to come.

The Current State of 3D Photonics

Over the past decade, 3D photonics has evolved rapidly, driven by breakthroughs in materials science, fabrication technologies, and computational modeling. These advancements have enabled the creation of complex photonic structures with unprecedented precision and performance. The integration of photonic devices into various applications, from telecommunications and data centers to medical diagnostics and environmental monitoring, has demonstrated the versatility and impact of this technology.

Co-packaged optics (CPO) represents a significant advancement in the design and performance of data centers and high-performance computing systems. By integrating optical components directly with electronic circuits in a single package, CPO drastically reduces electrical power consumption and latency associated with traditional interconnects. This integration minimizes the distance between high-speed electrical and optical signals, enhancing signal integrity and enabling higher data rates over longer distances. The adoption of CPO is driven by the growing demand for increased bandwidth and efficiency in data centers, which support a wide range of applications from cloud computing to artificial intelligence. As the industry moves towards higher data throughput and lower energy consumption, co-packaged optics is poised to play a crucial role in meeting these challenges, providing a scalable and efficient solution for future networking needs.

The progress in 3D photonics has been fueled by interdisciplinary collaboration, bringing together experts from physics, engineering, materials science, computer science, and other fields. This collaborative approach has facilitated the development of innovative solutions and accelerated the translation of research into practical applications. The synergy between academia, industry, and government agencies has been particularly instrumental in advancing photonic technologies, providing the necessary resources and infrastructure for groundbreaking research and development.

Market adoption of 3D photonics is on the rise, with significant investments from leading technology companies and increasing demand across various sectors. The benefits of photonic technologies, such as higher data transmission speeds, enhanced imaging capabilities, and improved sensing accuracy, are driving their adoption in critical applications. However, challenges such as high production costs, material limitations, and complex fabrication processes remain. Addressing these challenges is essential for the widespread commercialization and deployment of 3D photonic devices.

Future Directions and Opportunities

The field of quantum photonics holds immense promise for the future, with the potential to revolutionize computing, communication, and cryptography. Advances in quantum photonic circuits, quantum key distribution, and quantum sensing are expected to unlock new levels of performance and security. The development of scalable and reliable quantum photonic systems will be a key focus in the coming years. Overcoming technical challenges related to qubit integration, error correction, and system stability will be crucial for realizing the full potential of quantum photonics.

The integration of artificial intelligence (AI) and machine learning (ML) with 3D photonics is poised to drive significant advancements. AI and ML algorithms can optimize the design and performance of photonic devices, enabling more efficient and intelligent systems. Applications such as autonomous vehicles, smart cities, and healthcare diagnostics will benefit from AI-driven photonic technologies. The combination of AI and photonics will enable real-time data processing, predictive analytics, and adaptive responses, enhancing the capabilities of these systems.

As global concerns about sustainability and environmental impact grow, 3D photonics offers solutions for creating more energy-efficient and environmentally friendly technologies. Photonic devices with lower power consumption and improved performance can contribute to reducing the carbon footprint of various industries. Innovations in photonic sensors for environmental monitoring, energy-efficient lighting, and sustainable manufacturing practices will play a crucial role in addressing environmental challenges. The development of circular economy principles in photonics will further enhance sustainability efforts.

The future of healthcare will be significantly influenced by advancements in 3D photonics. Enhanced imaging techniques, non-invasive diagnostic tools, and photonic therapeutic devices will revolutionize medical diagnostics and treatment. Personalized medicine, enabled by real-time monitoring and precise imaging, will become more prevalent. Photonic technologies will facilitate early detection of diseases, improve treatment outcomes, and reduce healthcare costs, ultimately enhancing patient care and quality of life.

The demand for high-speed data transmission and efficient data centers will continue to drive innovations in 3D photonics. The deployment of silicon photonics and advanced optical interconnects will enable faster, more reliable communication networks. The evolution of 5G and the emergence of 6G networks will rely heavily on photonic technologies to meet increasing data demands. Photonic integration will support the growth of cloud computing, edge computing, and IoT, transforming the telecommunications landscape.

Challenges and Solutions

Despite the progress made, several technical challenges persist in the field of 3D photonics. These include complex fabrication processes, material limitations, thermal management, and ensuring high yield and quality control in production. Addressing these challenges will require continued investment in research and development, interdisciplinary collaboration, and the adoption of advanced manufacturing techniques. Exploring new materials, improving simulation and modeling tools, and developing scalable production methods will be key to overcoming these obstacles.

The path to market adoption and commercialization of 3D photonic technologies can be fraught with challenges. High production costs, regulatory and compliance issues, and market uncertainty can hinder the widespread deployment of photonic devices. Strategies to enhance market adoption include conducting thorough market research, engaging with potential customers early in the development process, and developing flexible and versatile products. Establishing strong industry-academia partnerships and participating in standardization efforts will also facilitate commercialization.

The growth of the 3D photonics industry depends on the availability of skilled talent and a well-trained workforce. Attracting and retaining top talent in this competitive field can be challenging. Investing in education and training programs, fostering a culture of continuous learning, and providing opportunities for professional growth will be crucial for building a skilled workforce. Encouraging diversity and inclusion in the field will also bring fresh perspectives and drive innovation.

Forward-Looking Statements

The field of 3D photonics is poised for continued innovation, with ongoing research and development expected to yield new breakthroughs and applications. The integration of photonics with emerging technologies such as AI, quantum computing, and IoT will drive further advancements. As researchers push the boundaries of what is possible with photonic technologies, we can expect to see new devices and systems that enhance our capabilities and improve our quality of life.

Global collaboration will be essential for advancing 3D photonics. International partnerships, collaborative research initiatives, and knowledge-sharing platforms will facilitate the exchange of ideas and resources, accelerating the pace of innovation. Addressing global challenges, such as climate change, healthcare, and communication, will require concerted efforts and collaboration across borders. The photonics community must work together to develop solutions that benefit society as a whole.

The societal impact of 3D photonics will be profound, with potential benefits in areas such as healthcare, environmental sustainability, and communication. Photonic technologies will play a key role in addressing some of the most pressing challenges of our time. As photonics continues to evolve, it will enable new opportunities for economic growth, improve quality of life, and contribute to a more sustainable and connected world (Fig. 12.2).

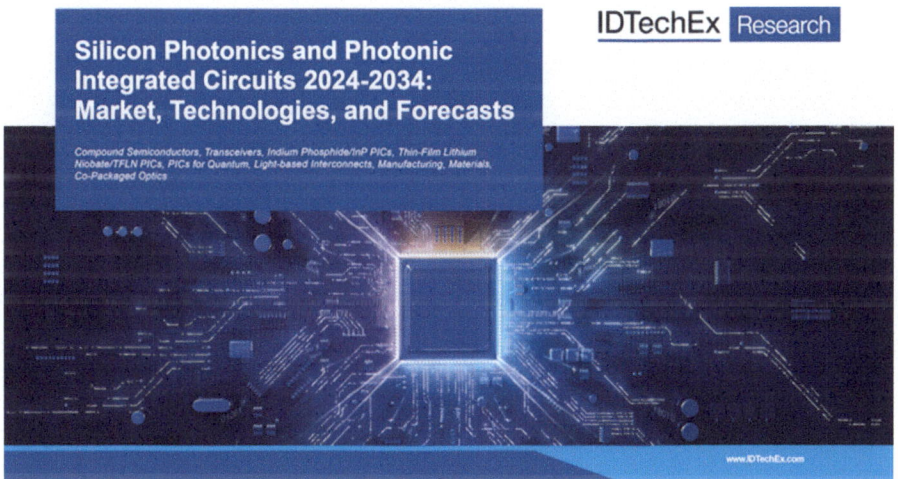

Fig. 12.2 Final roadmap for the future of 3D photonics, showing anticipated milestones in the field over the next decade. https://www.idtechex.com/en/research-report/silicon-photonics-and-photonic-integrated-circuits-2024-2034-market-technologies-and-forecasts/1006

Summary

The landscape of 3D photonics is indeed a dynamic and rapidly evolving field, characterized by interdisciplinary collaboration and a remarkable potential to transform technology across multiple sectors. The insights and predictions in this book offer a detailed view of both the current advancements and the future possibilities that 3D photonics holds. As technologies in telecommunications, healthcare, quantum computing, and environmental monitoring continue to demand higher efficiency, faster data processing, and compact design, the unique capabilities of 3D photonics to integrate complex photonic elements into multi-layered structures become increasingly invaluable. Realizing the full potential of 3D photonics will require the dedication of a diverse range of researchers, engineers, and industry leaders who bring expertise from fields such as materials science, optical engineering, and computational modeling. The future of 3D photonics promises groundbreaking developments that will not only push the boundaries of current technologies but also introduce entirely new applications, shaping the future of technology and society in profound ways. This field stands as a beacon of immense promise and opportunity, poised to drive innovation and solve critical challenges in the decades to come.

Bibliography

1. Kuyken, B. et al. 50 dB parametric on-chip gain in silicon photonic wires. *Opt. Lett. 36*, 4401–4403 (2011)
2. Lamont, M. R. et al. Net-gain from a parametric amplifier on a chalcogenide optical chip. *Opt. Express 16*, 20374–20381 (2008)
3. Foster, M. A. et al. Broad-band optical parametric gain on a silicon photonic chip. *Nature 441*, 960–963 (2006)
4. Morichetti, F. et al. Travelling-wave resonant four-wave mixing breaks the limits of cavity-enhanced all-optical wavelength conversion. *Nat. Commun. 2*, 296 (2011)
5. Wang, K.-Y. & Foster, A. C. GHz-rate optical parametric amplifier in hydrogenated amorphous silicon. *J. Opt. 17*, 094012 (2015)
6. Pu, M. et al. Ultra-efficient and broadband nonlinear AlGaAs-on-insulator chip for low-power optical signal processing. *Laser Photonics Rev. 12*, 1800111 (2018)
7. Ooi, K. et al. Pushing the limits of CMOS optical parametric amplifiers with USRN:Si_7N_3 above the two-photon absorption edge. *Nat. Commun. 8*, 13878 (2017)
8. Liu, X., Osgood, R. M., Vlasov, Y. A. & Green, W. M. Mid-infrared optical parametric amplifier using silicon nanophotonic waveguides. *Nat. Photon. 4*, 557–560 (2010)
9. Gajda, A. et al. Highly efficient CW parametric conversion at 1550 nm in SOI waveguides by reverse biased p-i-n junction. *Opt. Express 20*, 13100–13107 (2012)
10. Umeki, T., Tadanaga, O., Takada, A. & Asobe, M. Phase sensitive degenerate parametric amplification using directly-bonded PPLN ridge waveguides. *Opt. Express 19*, 6326–6332 (2011)
11. Kishimoto, T., Inafune, K., Ogawa, Y., Sasaki, H. & Murai, H. Highly efficient phase-sensitive parametric gain in periodically poled $LiNbO_3$ ridge waveguide. *Opt. Lett. 41*, 1905–1908 (2016)
12. Hansryd, J. & Andrekson, P. A. Broad-band continuous-wave-pumped fiber optical parametric amplifier with 49-dB gain and wavelength-conversion efficiency. *IEEE Photonics Technol. Lett. 13*, 194–196 (2001)
13. Marhic, M., Kagi, N., Chiang, T.-K. & Kazovsky, L. Broadband fiber optical parametric amplifiers. *Opt. Lett. 21*, 573–575 (1996)
14. Torounidis, T., Andrekson, P. A. & Olsson, B.-E. Fiber-optical parametric amplifier with 70-db gain. *IEEE Photonics Technol. Lett. 18*, 1194–1196 (2006)
15. Del'Haye, P., Arcizet, O., Gorodetsky, M. L., Holzwarth, R. & Kippenberg, T. J. Frequency comb assisted diode laser spectroscopy for measurement of microcavity dispersion. *Nat. Photon. 3*, 529–533 (2009)

16. Liu, J. et al. Frequency-comb-assisted broadband precision spectroscopy with cascaded diode lasers. Opt. Lett. 41, 3134–3137 (2016)
17. Zhoufeng Z, Feng C, Zhao Z et al (2020) Electronic-photonic arithmetic logic unit for high-speed computing. Nat Commun 11:2154
18. Dutta P, Bandyopadhyay C, Giri C et al (2014) Mach–Zehnder interferometer based all optical reversible carry-lookahead adder. In: Proceedings of the IEEE ISVLSI, pp 412–417
19. Lumerical Inc. Photonic integrated circuit simulator. https://www.lumerical.com/products/interconnect/ (Online). Accessed 17 Aug 2020
20. Luceda Phtotonics. IPKISS parametric design framework. http://www.lucedaphotonics.com (Online). Accessed 17 Aug 2020
21. Jin J (2014) The finite element method in electromagnetics. Wiley-IEEE Press, Hoboken
22. Pedrola GL (2015) Beam propagation method for design of optical waveguide devices. Wiley, Hoboken
23. Condrat C, Kalla P, Blair S (2011) Logic synthesis for integrated optics. In: Proceedings of the Great Lakes Symposium on Great Lakes Symposium on VLSI (GLSVLSI), pp 13–18
24. Deb A, Wille R, Keszöcze O et al (2017) Synthesis of optical circuits using binary decision diagrams. Integration 59(C):42–51
25. Bandyopadhyay C, Das R, Wille R et al (2018) Synthesis of circuits based on all-optical Mach–Zehnder interferometers using binary decision diagrams. Microelectron J 71:19–29
26. Das R, Bandyopadhyay C, Rahaman H (2016) All optical reversible design of Mach–Zehnder interferometer based carry-skip adder. In: Proceedings of the IEEE DISCOVER, pp 73–78
27. Roy JN, Chattopadhyay T, Sarkar T (2010) All-optical multiplication using SOA-MZI based programmable logic device (PLD). In: Proceedings of the International Conference on Communication, Computers and Devices
28. Nivedita, Kaur S, Goyal R (2019) All-optical decoder/demultiplexer with enable using SOA based Mach–Zehnder Interferometers. In: Proceedings of the International Conference on Signal Processing and Integrated Networks (SPIN), pp 780–784
29. Dutta P, Bandyopadhyay C, Rahaman H (2014) All optical implementation of Mach–Zehnder interferometer based reversible sequential circuit. In: Proceedings of the International Symposium on VLSI Design and Test, pp 1–2
30. Dutta P, Bandyopadhyay C, Rahaman H. (2015) All optical implementation of Mach–Zehnder interferometer based reversible sequential counters. In: Proceedings of the International Conference on VLSI Design (VLSID), pp 232–237
31. Papadimitriou GI, Papazoglou C, Pomportsis AS (2003) Optical switching: switch fabrics, techniques, and architectures. IEEE J Lightw Technol 21(2):384–405
32. Testa F, Pavesi L (2018) Optical switching in next generation data centers. Springer, Berlin
33. Kostinski N, Fok MP, Prucnal PR (2009) Experimental demonstration of an all-optical fiber-based Fredkin gate. OSA Opt Lett 34(18):2766–2768
34. Kumar S, Chattopadhyay T, Sengupta I (2015) Implementation of 2-bit multiplier based on electro-optic effect in Mach–Zehnder interferometers. Opt Quantum Electron 47(12):3667–3688
35. Wang Q, Zhu G, Chen H et al (2004) Study of all-optical XOR using Mach–Zehnder interferometer and differential scheme. EEE J Quantum Electron 40(6):703–710
36. Earnshaw MP, Cappuzzo MA, Chen E et al (2007) Ultra-low power thermo-optic silica-on-silicon waveguide membrane switch. IET Electron Lett 43(7):393–394
37. Law FK, Uddin M, Chen AC et al (2020) Positive edge-triggered JK flip-flop using silicon-based micro-ring resonator. Opt Quantum Electron 52:314:1314:1–314:12314:12
38. Taraphdar C, Chattopadhyay T, Roy JN (2010) Mach–Zehnder interferometer-based all-optical reversible logic gate. Opt Laser Technol 42(2):249–259

39. Manna A, Saha S, Das R et al (2017) All optical design of cost efficient multiplier circuit using terahertz optical asymmetric demultiplexer. In: Proceedings of the IEEE/ACM International Conference on Computer-Aided Design (ICCAD), pp 1–5
40. Kumar A, Medhekar S (2020) All optical NOR and NAND gates using four circular cavities created in 2D nonlinear photonic crystal. Opt Laser Technol 123:105910:1–105910:9
41. Cheng Q, Rumley S, Bahadori M et al (2018) Photonic switching in high performance datacenters [Invited]. OSA Opt Express 26(12):16022–16043
42. Tu X, Song C, Huang T et al (2019) Implementation of reversible Peres gate using electro-optic effect inside lithium-niobate based Mach–Zehnder interferometers. MDPI Micromach 10:1
43. Dong H, Sun H, Wang Q et al (2006) 80 Gb/s all-optical logic AND operation using Mach–Zehnder interferometer with differential scheme. Opt Commun 265(1):79–83
44. Xu M, Wang Y, Zhai M et al (2018) Ultrafast passive all-optical full function logic gates on micro-silicon-on-insulator waveguide chip. SPIE J Nanophotonics 12(4):1–17
45. Datta K, Chattopadhyay T, Sengupta I (2015) All optical design of binary adders using semiconductor optical amplifier assisted Mach–Zehnder interferometer. Microelectron J 46(9):839–847
46. Ying Z, Wang Z, Zhao Z et al (2018) Microdisk-based full adders for optical computing in silicon photonics. In: Proceedings of the IEEE Conference on Lasers and Electro-optics (CLEO), pp 1–2
47. Ying Z, Wang Z, Zhao Z et al (2018) Silicon microdisk-based full adders for optical computing. OSA Opt Express 43(5):983–986
48. Pal A, Kumar S, Sharma S (2018) Design of optical decoder circuits using electro-optic effect inside Mach–Zehnder interferometers for high speed communication. Photonic Netw Commun 35(1):79–89
49. Datta K, Sengupta I (2014) All optical reversible multiplexer design using Mach–Zehnder interferometer. In: Proceedings of the International Conference on VLSI Design (VLSID), pp 539–544
50. Sharma S, Chakrabarty K, Roy S (2018) On designing all-optical multipliers using Mach–Zender interferometers. In: Proceedings of the Euromicro Conference on Digital System Design (DSD), pp 672–679
51. Kumar A, Kumar S, Raghuwanshi SK (2014) Implementation of full-adder and full-subtractor based on electro-optic effect in Mach–Zehnder interferometers. Opt Commun 324(21):97–107
52. Chattopadhyay T, Roy JN (2011) All-optical method of developing half and full subtractors by the use of phase encoding principle. Opt Int J Light Electron Opt 122(24):2207–2210
53. Abdulnabi SH, Abbas MN (2019) Design an all-optical combinational logic circuits based on nano-ring insulator-metal-insulator plasmonic waveguides. MDPI Photonics 6(1):1–13
54. Bhattacharyya A, Gayen DK, Chattopadhyay T (2016) Design of 2-to-4 all-optical decoder with the help of terahertz optical asymmetric demultiplexer. Int J Mod Nonlinear Theory Appl 22(5):67–72
55. Kaler R, Kaler RS (2011) Implentation of optical encoder and multiplexer using Mach–Zehnder inferometer. Opt Int J Light Electron Opt 122(15):1399–1405
56. Rajasekar R, Latha R, Robinson S (2019) Ultra-contrast ratio optical encoder using photonic crystal waveguide. Mater Lett 251(5):144–147
57. Chattopadhyay T, Roy JN (2011) Semiconductor optical amplifier (SOA)-assisted Sagnac switch for designing of all-optical tri-state logic gates. Opt Int J Light Electron Opt 122(12):1073–1078
58. Kim JH, Kim BC, Byun YT et al (2004) All-optical AND gate using cross-gain modulation in semiconductor optical amplifiers. Jpn J Appl Phys 43(2):608–610

59. Berrettini G, Simi A, Malacarne A et al (2006) Ultrafast integrable and reconfigurable XNOR, AND, NOR, and NOT photonic logic gate. IEEE Photonics Technol Lett 18(8):917–919
60. Singh P, Tripathi DK, Jaiswal S et al (2014) Designs of all-optical buffer and OR gate using SOA-MZI. Opt Quantum Electron 46(11):1435–1444
61. Singh P, Tripathi DK, Dixit HK (2014) Designs of all-optical NOR gates using SOA based MZI. Opt Int J Light Electron Opt 125(16):4437–4440
62. Koteb A (2017) Theoretical analysis of soliton NOR gate with semiconductor optical amplifier-assisted Mach–Zehnder interferometer. Opt Quantum Electron 49(5):1–11
63. Kotiyal S, Thapliyal H, Ranganathan H (2012) Mach–Zehnder interferometer based design of all optical reversible binary adder. In: Proceedings of the Design, Automation Test in Europe Conference (DATE), pp 721–726
64. Wille R, Keszöcze O, Hopfmuller C et al (2015) Reverse BDD-based synthesis for splitter-free optical circuits. In: Proceedings of the Asia and South Pacific Design Automation Conference (ASPDAC), pp 172–175
65. Chuang YK, Chen KJ, Lin KL et al (2018) PlanarONoC: concurrent placement and routing considering crossing minimization for optical networks-on-chip. In: Proceedings of the ACM/IEEE Design Automation Conference (DAC), pp 151:1–151:6
66. Liu D, Zhao Z, Wang Z et al (2018) OPERON: optical-electrical power-efficient route synthesis for on-chip signals. In: Proceedings of the ACM/IEEE Design Automation Conference (DAC), pp 1–6
67. Phoenix Software BV. Photonic design automation software. http://www.phoenixbv.com/ (Online). Accessed 17 Aug 2020
68. Heck, M. J. R. Highly integrated optical phased arrays: photonic integrated circuits for optical beam shaping and beam steering. *Nanophotonics 6*, 93–107 (2017)
69. Dostart, N. et al. Serpentine optical phased arrays for scalable integrated photonic lidar beam steering. *Optica 7*, 726 (2020)
70. Sun, J., Timurdogan, E., Yaacobi, A., Hosseini, E. S. & Watts, M. R. Large-scale nanophotonic phased array. *Nature 493*, 195–199 (2013)
71. Van Acoleyen, K., Bogaerts, W. & Baets, R. Two-dimensional dispersive off-chip beam scanner fabricated on silicon-on-insulator. *IEEE Photon. Technol. Lett. 23*, 1270–1272 (2011)
72. Fukui, T. et al. Non-redundant optical phased array. *Optica 8*, 1350 (2021)
73. Aflatouni, F., Abiri, B., Rekhi, A. & Hajimiri, A. Nanophotonic projection system. *Opt. Express 23*, 21012–21022 (2015)
74. Miller, S. A. et al. Large-scale optical phased array using a low-power multi-pass silicon photonic platform. *Optica 7*, 3 (2020)
75. Li, Y. et al. Wide-steering-angle high-resolution optical phased array. *Photon. Res. 9*, 2511 (2021)
76. Zadka, M. et al. On-chip platform for a phased array with minimal beam divergence and wide field-of-view. *Opt. Express 26*, 2528–2534 (2018)
77. Komljenovic, T., Helkey, R., Coldren, L. & Bowers, J. E. Sparse aperiodic arrays for optical beam forming and LIDAR. *Opt. Express 25*, 2511–2528 (2017)
78. Komljenovic, T. & Pintus, P. On-chip calibration and control of optical phased arrays. *Opt. Express 26*, 3199–3210 (2018)
79. Tran, M. A. et al. Extending the spectrum of fully integrated photonics to submicrometre wavelengths. *Nature 610*, 54–60 (2022)
80. Lin, Y. et al. Low-loss broadband bi-layer edge couplers for visible light. *Opt. Express 29*, 34565–34576 (2021)
81. Aslan, O. B. et al. Probing the optical properties and strain-tuning of ultrathin $Mo_{1-x}W_xTe_2$. *Nano Lett. 18*, 2485–2491 (2018)

82. Conley, H. J. et al. Bandgap engineering of strained monolayer and bilayer MoS_2. *Nano Lett.* *13*, 3626–3630 (2013)
83. Frisenda, R. et al. Biaxial strain tuning of the optical properties of single-layer transition metal dichalcogenides, *npj 2D Mater. Appl. 1*, 10 (2017)
84. Desai, S. B. et al. Strain-induced indirect to direct bandgap transition in multilayer WSe_2. *Nano Lett. 14*, 4592–4597 (2014)
85. Melitz, W., Shen, J., Kummel, A. C. & Lee, S. Surface science reports kelvin probe force microscopy and its application. *Surf. Sci. Rep. 66*, 1–27 (2011)
86. Kippenberg, T. J., Gaeta, A. L., Lipson, M. & Gorodetsky, M. L. Dissipative Kerr solitons in optical microresonators. *Science 361*, eaan8083 (2018)
87. Liang, W. et al. High spectral purity Kerr frequency comb radio frequency photonic oscillator. *Nat. Commun. 6*, 8957 (2015)
88. Yang, K. Y. et al. Bridging ultrahigh-Q devices and photonic circuits. *Nat. Photon. 12*, 297–302 (2018)
89. Morton, P. A. & Morton, M. J. High-power, ultra-low noise hybrid lasers for microwave photonics and optical sensing. *J. Lightw. Technol. 36*, 5048–5057 (2018)
90. Huang, D. et al. High-power sub-kHz linewidth lasers fully integrated on silicon. *Optica 6*, 745–752 (2019)
91. Tian, H. et al. Hybrid integrated photonics using bulk acoustic resonators. Preprint at https://arxiv.org/abs/1907.10177 (2019)
92. Joshi, C. et al. Thermally controlled comb generation and soliton modelocking in microresonators. *Opt. Lett. 41*, 2565–2568 (2016)
93. Mazur, M. et al. Enabling high spectral efficiency coherent superchannel transmission with soliton microcombs. Preprint at https://arxiv.org/abs/1812.11046 (2018)
94. Jouppi, N. P. et al. In-datacenter performance analysis of a tensor processing unit. In *Proc of the 44th Annual International Symposium on Computer Architecture* 1–12 (Association for Computing Machinery, 2017)
95. Watkins, D. S. *Fundamentals of Matrix Computations* 2ndedn, 640 (Wiley, 2002)
96. Palmer, T. Modelling: Build imprecise supercomputers. *Nature 526*, 32–33 (2015)
97. Altrock, P. M. et al. The mathematics of cancer: integrating quantitative models. *Nat. Rev. Cancer 15*, 730–745 (2015)
98. Achdou, Y. et al. Partial differential equation models in macroeconomics. *Philos. Trans. R. Soc. A Math. Phys. Eng. Sci. 372*, 20130397 (2014)
99. Bauer, P. et al. The quiet revolution of numerical weather prediction. *Nature 525*, 47–55 (2015)
100. Montgomery, D. C. et al. *Introduction to Linear Regression Analysis* 6th edn, 704 (John Wiley & Sons, 2021)
101. Hashima, S. & Muta, O. Fast matrix inversion methods based on Chebyshev and Newton iterations for zero forcing precoding in massive MIMO systems. *J. Wireless Com. Netw. 2020*, 34 (2020)
102. Rosário, F. et al. Fast matrix inversion updates for massive MIMO detection and precoding. *IEEE Signal Process. Lett. 23*, 75–79 (2016)
103. Prabhu, H. et al. Hardware efficient approximative matrix inversion for linear pre-coding in massive MIMO. in *2014 IEEE International Symposium on Circuits and Systems (ISCAS)* 1700–1703 (2014)
104. Tang, C. et al. High precision low complexity matrix inversion based on Newton iteration for data detection in the massive MIMO. *IEEE Commun. Lett. 20*, 490–493 (2016)
105. Zhang, C. et al. On the low-complexity, hardware-friendly tridiagonal matrix inversion for correlated massive MIMO systems. *IEEE Trans. Veh. Technol. 68*, 6272–6285 (2019)

106. Tanaka, G. et al. Recent advances in physical reservoir computing: a review. *Neural Netw. 115*, 100–123 (2019)
107. Lukoševičius, M. & Jaeger, H. Reservoir computing approaches to recurrent neural network training. *Comput. Sci. Rev. 3*, 127–149 (2009)
108. Shen, Y. et al. Deep learning with coherent nanophotonic circuits. *Nat. Photon. 11*, 441–446 (2017)
109. Feldmann, J. et al. Parallel convolutional processing using an integrated photonic tensor core. *Nature 589*, 52–58 (2021)
110. Zhou, H. et al. Photonic matrix multiplication lights up photonic accelerator and beyond. *Light Sci. Appl. 11*, 30 (2022)
111. Zhang, H. et al. An optical neural chip for implementing complex-valued neural network. *Nat. Commun. 12*, 457 (2021)
112. Vangeleyn, M., Griffin, P. F., Riis, E. & Arnold, A. S. Laser cooling with a single laser beam and a planar diffractor. *Opt. Lett. 35*, 3453–3455 (2010)
113. Nshii, C. C. et al. A surface-patterned chip as a strong source of ultracold atoms for quantum technologies. *Nat. Nanotechnol. 8*, 321–324 (2013)
114. Lee, J., Grover, J. A., Orozco, L. A. & Rolston, S. L. Sub-Doppler cooling of neutral atoms in a grating magneto-optical trap. *J. Opt. Soc. Am. B 30*, 2869–2874 (2013)
115. Imhof, E. et al. Two-dimensional grating magneto-optical trap. *Phys. Rev. A 96*, 033636 (2017)
116. McGilligan, J. P. et al. Grating chips for quantum technologies. *Sci. Rep. 7*, 384 (2017)
117. Barker, D. S. et al. Single-beam Zeeman slower and magneto-optical trap using a nanofabricated grating. *Phys. Rev. Appl. 11*, 064023 (2019)
118. Sitaram, A. et al. Confinement of an alkaline-earth element in a grating magneto-optical trap. *Rev. Sci. Instrum. 91*, 103202 (2020)
119. McGehee, W. R. et al. Magneto-optical trapping using planar optics. *New J. Phys. 23*, 013021 (2021)
120. Seo, S. et al. Maximized atom number for a grating magneto-optical trap via machine-learning assisted parameter optimization. *Opt. Express 29*, 35623–35639 (2021)
121. Soh, D. B. S., Biedermann, G., Lee, J. & Schwindt, P. Modeling of atom interferometer accelerometer. *SAND Rep. 2020*, 10087 (2020)
122. Nevlacsil, S. *et al.* Multi-channel swept source optical coherence tomography concept based on photonic integrated circuits. *Opt. Express 28*, 32468–32482. https://doi.org/10.1364/OE.404588 (2020)
123. Errando-Herranz, C. et al. MEMS for photonic integrated circuits. *IEEE J. Sel. Topics Quantum Electron. 26*, 1–16 (2020)
124. Gyger, S. et al. Reconfigurable photonics with on-chip single-photon detectors. *Nat. Commun. 12*, 1408 (2021)
125. Stanfield, P. R., Leenheer, A. J., Michael, C. P., Sims, R. & Eichenfield, M. CMOS-compatible, piezo-optomechanically tunable photonics for visible wavelengths and cryogenic temperatures. *Opt. Express 27*, 28588–28605 (2019)
126. Awschalom, D. D., Hanson, R., Wrachtrup, J. & Zhou, B. B. Quantum technologies with optically interfaced solid-state spins. *Nat. Photon. 12*, 516–527 (2018)
127. Atatüre, M., Englund, D., Vamivakas, N., Lee, S.-Y. & Wrachtrup, J. Material platforms for spin-based photonic quantum technologies. *Nat. Rev. Mater. 3*, 38–51 (2018)
128. Humphreys, P. C. et al. Deterministic delivery of remote entanglement on a quantum network. *Nature 558*, 268–273 (2018); correction *562*, E2 (2018)
129. Bradley, C. E. et al. A ten-qubit solid-state spin register with quantum memory up to one minute. *Phys. Rev. X 9*, 031045 (2019)

130. Bhaskar, M. K. et al. Experimental demonstration of memory-enhanced quantum communication. *Nature 580*, 60–64 (2020)
131. Muralidharan, S. et al. Optimal architectures for long distance quantum communication. *Sci. Rep. 6*, 20463 (2016)
132. Lo Piparo, N., Munro, W. J. & Nemoto, K. Quantum multiplexing. *Phys. Rev. A 99*, 022337 (2019)
133. Nemoto, K. et al. Photonic architecture for scalable quantum information processing in diamond. *Phys. Rev. X 4*, 031022 (2014)
134. Monroe, C. et al. Large-scale modular quantum-computer architecture with atomic memory and photonic interconnects. *Phys. Rev. A 89*, 022317 (2014)
135. Nickerson, N. H., Fitzsimons, J. F. & Benjamin, S. C. Freely scalable quantum technologies using cells of 5-to-50 qubits with very lossy and noisy photonic links. *Phys. Rev. X 4*, 041041 (2014)
136. Choi, H., Pant, M., Guha, S. & Englund, D. Percolation-based architecture for cluster state creation using photon-mediated entanglement between atomic memories. *npj Quantum Inf. 5*, 104 (2019)
137. Lu, T.-J. et al. Aluminum nitride integrated photonics platform for the ultraviolet to visible spectrum. *Opt. Express 26*, 11147–11160 (2018)
138. Atabaki, A. H. et al. Integrating photonics with silicon nanoelectronics for the next generation of systems on a chip. *Nature 556*, 349–354 (2018); *560*, E4 (2018)
139. Harris, N. C. et al. Linear programmable nanophotonic processors. *Optica 5*, 1623–1631 (2018)
140. Blain, M. G. et al. Hybrid MEMS-CMOS ion traps for NISQ computing. *Quantum Sci. and Technol. 6*, 034011 (2021)
141. Sutherland, R. T. et al. Laser-free trapped-ion entangling gates with simultaneous insensitivity to qubit and motional decoherence. *Phys. Rev. A 101*, 042334 (2020)
142. Siegele-Brown, M. et al. Fabrication of surface ion traps with integrated current carrying wires enabling high magnetic field gradients. *Quant. Sci. Technol. 7*, 034003 (2022)
143. Mehta, K. K. et al. Integrated optical multi-ion quantum logic. *Nature 586*, 533–537 (2020)
144. Pérez, D. et al. Multipurpose silicon photonics signal processor core. *Nat. Commun. 8*, 1–9 (2017)
145. Zhang, W. & Yao, J. Photonic integrated field-programmable disk array signal processor. *Nat. Commun. 11*, 406 (2020)
146. Reed, G. T., Mashanovich, G., Gardes, F. Y. & Thomson, D. J. Silicon optical modulators. *Nat. Photon 4*, 518–526 (2010)
147. Wang, C. et al. Integrated lithium niobate electro-optic modulators operating at CMOS-compatible voltages. *Nature 562*, 101–104 (2018)
148. Wuttig, M., Bhaskaran, H. & Taubner, T. Phase-change materials for non-volatile photonic applications. *Nat. Photonics 11*, 465–476 (2017)
149. Abdollahramezani, S. et al. Tunable nanophotonics enabled by chalcogenide phase-change materials. *Nanophotonics 9*, 1189–1241 (2020)
150. Fang, Z., Chen, R., Zheng, J. & Majumdar, A. Non-volatile reconfigurable silicon photonics based on phase-change materials. *IEEE J. Sel. Top. Quantum Electron. 28*, 8200317 (2022)
151. Chen, R. et al. Broadband nonvolatile electrically controlled programmable units in silicon photonics. *ACS Photonics 9*, 2142–2150 (2022)
152. Zheng, J. et al. Nonvolatile electrically reconfigurable integrated photonic switch enabled by a silicon PIN diode heater. *Adv. Mater. 32*, 2001218 (2020)
153. Ríos, C. et al. Ultra-compact nonvolatile phase shifter based on electrically reprogrammable transparent phase change materials. *PhotoniX 3*, 26 (2022)

154. Fang, Z. et al. Ultra-low-energy programmable non-volatile silicon photonics based on phase-change materials with graphene heaters. *Nat. Nanotechnol. 17*, 842–848 (2022)
155. Dong, W. et al. Wide bandgap phase change material tuned visible photonics. *Adv. Funct. Mater. 29*, 1806181 (2019)
156. Jackson, Z. et al. Automated Manufacture of Autologous CD19 CAR-T Cells for Treatment of Non-Hodgkin Lymphoma. *Front. Immunol. 11*, (2020)
157. Yang, J. et al. Next-day manufacture of a novel anti-CD19 CAR-T therapy for B-cell acute lymphoblastic leukemia: first-in-human clinical study. *Blood Cancer J. 12*, 104 (2022)
158. de Wijs, K. et al. Micro vapor bubble jet flow for safe and high-rate fluorescence-activated cell sorting. *Lab Chip 17*, 1287–1296 (2017)
159. Butement, J. T. et al. Monolithically-integrated cytometer for measuring particle diameter in the extracellular vesicle size range using multi-angle scattering. *Lab Chip 20*, 1267–1280 (2020)
160. Friis, P. et al. Monolithic integration of microfluidic channels and optical waveguides in silica on silicon. *Appl. Opt. 40*, 6246 (2001)
161. Watts, B. R., Zhang, Z., Xu, C. Q., Cao, X. & Lin, M. A photonic-microfluidic integrated device for reliable fluorescence detection and counting. *Electrophoresis 33*, 3236–3244 (2012)
162. Butement, J. T. et al. Integrated optical waveguides and inertial focussing microfluidics in silica for microflow cytometry applications. *J. Micromech. Microeng. 26*, (2016)
163. Tabatabaei Mashayekh, A. et al. Multi-color flow cytometer with PIC-based structured illumination. *Biomed. Opt. Express. 13*, (2022)
164. Verellen, N. et al. Integrated photonics for miniature flow cytometry. in (2019). https://doi.org/10.7567/ssdm.2017.f-4-01
165. Kerman, S. et al. Integrated nanophotonic excitation and detection of fluorescent microparticles. *ACS Photonics 4*, (2017)
166. Zhao, Y., Li, Q. & Hu, X. Universally applicable three-dimensional hydrodynamic focusing in a single-layer channel for single cell analysis. *Anal. Methods 10* (2018)
167. Haindl, R. et al. Ultra-high-resolution SD-OCM imaging with a compact polarization-aligned 840 nm broadband combined-SLED source. *Biomed. Opt. Express 11*, 3395–3406 (2020)
168. De Boer, J. F., Leitgeb, R. & Wojtkowski, M. Twenty-five years of optical coherence tomography: the paradigm shift in sensitivity and speed provided by Fourier domain OCT [Invited]. *Biomed. Opt. Express 8*, 3248–3280 (2017)
169. Schneider, S. et al. Optical coherence tomography system mass-producible on a silicon photonic chip. *Opt. Express. 24*, 1573–1586. https://doi.org/10.1364/OE.24.001573 (2016) (*Publisher: OSA*)
170. Eggleston, M. S. et al. 90 dB sensitivity in a chip-scale swept-source optical coherence tomography system. *Conference on Lasers and Electro-Optics.* (Optical Society of America, San Jose, 2018)
171. Zhu, D. et al. Integrated photonics on thin-film lithium niobate. *Adv. Opt. Photonics 13*, 242–352 (2021)
172. Boes, A. et al. Lithium niobate photonics: unlocking the electromagnetic spectrum. *Science 379*, eabj4396 (2023)
173. Butaud, E. et al. Innovative Smart Cut piezo on insulator (POI) substrates for 5G acoustic filters. In *2020 IEEE International Electron Devices Meeting (IEDM)* (ed. Datta, S.) 34.6.1–34.6.4 (IEEE, 2020)
174. Li, Z. et al. High density lithium niobate photonic integrated circuits. *Nat. Commun. 14*, 4856 (2023)

175. Ballandras, S. et al. New generation of SAW devices on advanced engineered substrates combining piezoelectric single crystals and silicon. In *2019 Joint Conference of the IEEE International Frequency Control Symposium and European Frequency and Time Forum (EFTF/IFC)* 1–6 (IEEE, 2019)
176. Yan, Y. et al. Wafer-scale fabrication of 42° rotated y-cut LiTaO$_3$-on-insulator (LTOI) substrate for a SAW resonator. *ACS Appl. Electron. Mater. 1*, 1660–1666 (2019)
177. SOITEC. Capital markets day 2021. https://www.soitec.com/en/capital-markets-day-2021 (2021)
178. Luke, K. et al. Wafer-scale low-loss lithium niobate photonic integrated circuits. *Opt. Express 28*, 24452–24458 (2020)
179. Shams-Ansari, A. et al. Reduced material loss in thin-film lithium niobate waveguides. *APL Photonics 7*, 081301 (2022)
180. Wang, J., Chen, P., Dai, D. & Liu, L. Polarization coupling of X-cut thin film lithium niobate based waveguides. *IEEE Photonics J.12*, 2200310 (2020)
181. Herr, T. et al. Temporal solitons in optical microresonators. *Nat. Photon. 8*, 145–152 (2014)
182. Zhang, M. et al. Broadband electro-optic frequency comb generation in a lithium niobate microring resonator. *Nature 568*, 373–377 (2019)
183. Hu, Y. et al. High-efficiency and broadband on-chip electro-optic frequency comb generators. *Nat. Photon. 16*, 679–685 (2022)
184. Han, Y. et al. Electrically pumped widely tunable O-band hybrid lithium niobite/III–V laser. *Opt. Lett. 46*, 5413–5416 (2021)
185. Op de Beeck, C. et al. III–V-on-lithium niobate amplifiers and lasers. *Optica 8*, 1288–1289 (2021)
186. Chang, L. et al. Heterogeneous integration of lithium niobate and silicon nitride waveguides for wafer-scale photonic integrated circuits on silicon. *Opt. Lett. 42*, 803–806 (2017)
187. Liu, J. et al. Monolithic piezoelectric control of soliton microcombs. *Nature 583*, 385–390 (2020)
188. Lihachev, G. et al. Low-noise frequency-agile photonic integrated lasers for coherent ranging. *Nat. Commun. 13*, 3522 (2022)
189. Churaev, M. et al. A heterogeneously integrated lithium niobate-on-silicon nitride photonic platform. Preprint at https://arxiv.org/abs/2112.02018 (2021)
190. Liu, J. et al. Double inverse nanotapers for efficient light coupling to integrated photonic devices. *Opt. Lett. 43*, 3200–3203 (2018)
191. Kondratiev, N. M. et al. Self-injection locking of a laser diode to a high-Q WGM microresonator. *Opt. Express 25*, 28167–28178 (2017)
192. Duthel, T. et al. Laser linewidth estimation by means of coherent detection. *IEEE Photon. Technol. Lett. 21*, 1568–1570 (2009)
193. Komljenovic, T. et al. Heterogeneous silicon photonic integrated circuits. *J. Light Technol. 34*, 20–35 (2016)
194. Jones, R. et al. Heterogeneously integrated InP/silicon photonics: fabricating fully functional transceivers. *IEEE Nanotechnol. Mag. 13*, 17–26 (2019)
195. Margalit, N. et al. Perspective on the future of silicon photonics and electronics. *Appl. Phys. Lett. 118*, 220501 (2021)
196. Morton, P. A. & Morton, M. J. High-power, ultra-low noise hybrid lasers for microwave photonics and optical sensing. *J. Light. Technol. 36*, 5048–5057 (2018)
197. Kippenberg, T. J., Holzwarth, R. & Diddams, S. A. Microresonator-based optical frequency combs. *Science 332*, 555–559 (2011)
198. Gaeta, A. L., Lipson, M. & Kippenberg, T. J. Photonic-chip-based frequency combs. *Nat. Photonics 13*, 158–169 (2019)

199. Jin, W. et al. Hertz-linewidth semiconductor lasers using CMOS-ready ultra-high-Q microresonators. *Nat. Photonics 15*, 346–353 (2021)
200. Maleki, L. et al. High performance, miniature hyper-parametric microwave photonic oscillator. In *Proc 2010 IEEE International Frequency Control Symposium*, 558–563 (IEEE, 2010)
201. Maleki, L. & Matsko, A. B. Generation of single optical tone, RF oscillation signal and optical comb in a triple-oscillator device based on nonlinear optical resonator (2014). US Patent 8,681, 827
202. Liang, W. et al. High spectral purity Kerr frequency comb radio frequency photonic oscillator. *Nat. Commun. 6*, 7957 (2015)
203. Liu, J. et al. Photonic microwave generation in the X- and K-band using integrated soliton microcombs. *Nat. Photonics 14*, 486–491 (2020)
204. Marin-Palomo, P. et al. Microresonator-based solitons for massively parallel coherent optical communications. *Nature 546*, 274 (2017)
205. Xiang, C., Morton, P. A. & Bowers, J. E. Ultra-narrow linewidth laser based on a semiconductor gain chip and extended Si_3N_4 Bragg grating. *Opt. Lett. 44*, 3825–3828 (2019)
206. Norman, J. C., Jung, D., Wan, Y. & Bowers, J. E. Perspective: the future of quantum dot photonic integrated circuits. *APL Photonics 3*, 030901 (2018)
207. He, Y. et al. Self-starting bi-chromatic $LiNbO_3$ soliton microcomb. *Optica 6*, 1138–1144 (2019)
208. He, M. et al. High-performance hybrid silicon and lithium niobate Mach–Zehnder modulators for 100 $Gbits^{-1}$ and beyond. *Nat. Photonics 13*, 359–364 (2019)
209. Diekmann, R. et al. Chip-based wide field-of-view nanoscopy. *Nat. Photon. 11*, 322–328 (2017)
210. Tinguely, J. C. et al. Photonic-chip assisted correlative light and electron microscopy. *Commun. Biol. 3*, 739 (2020)
211. Opstad, I. S. et al. Fluorescence fluctuation-based super-resolution microscopy using multimodal waveguided illumination. *Opt. Express 29*, 23368–23380 (2021)
212. Archetti, A. et al. Waveguide-PAINT offers an open platform for large field-of-view super-resolution imaging. *Nat. Commun. 10*, 1267 (2019)
213. Ojaghi, A. et al. Label-free hematology analysis using deep-ultraviolet microscopy. *Proc. Natl Acad. Sci. USA 117*, 14779–14789 (2020)
214. Geim, A. K. Graphene: status and prospects. *Science 324*, 1530–1534 (2009)
215. Peyskens, F. et al. Integration of single photon emitters in 2D layered materials with a silicon nitride photonic chip. *Nat. Commun. 10*, 4435 (2019)
216. Genco, A. et al. Optical nonlinearity goes ultrafast in 2D semiconductor-based nanocavities. *Light Sci. Appl. 11*, 127 (2022)
217. Yin, X. et al. Edge nonlinear optics on a MoS_2 atomic monolayer. *Science 344*, 488–490 (2014)
218. Shree, S. et al. Interlayer exciton mediated second harmonic generation in bilayer MoS_2. *Nat. Commun. 12*, 6894 (2021)
219. Ye, Y. et al. Monolayer excitonic laser. *Nat. Photon 9*, 733–737 (2015)
220. Wu, S. et al. Monolayer semiconductor nanocavity lasers with ultralow thresholds. *Nature 520*, 69–72 (2015)
221. Chen, H. et al. Enhanced second-harmonic generation from two-dimensional $MoSe_2$ on a silicon waveguide. *Light Sci. Appl. 6*, e17060–e17060 (2017)
222. Liu, N. et al. Silicon nitride waveguides with directly grown WS_2 for efficient second-harmonic generation. *Nanoscale 14*, 49–54 (2022)
223. Yang, S. et al. CMOS-compatible WS_2-based all-optical modulator. *Acs Photon 5*, 342–346 (2018)
224. Sun, Z. et al. Optical modulators with 2D layered materials. *Nat. Photon. 10*, 227–238 (2016)

225. Flöry, N. et al. Waveguide-integrated van der Waals heterostructure photodetector at telecom wavelengths with high speed and high responsivity. *Nat. Nanotechnol.* 15, 118–124 (2020)
226. Li, F. et al. Recent progress of silicon integrated light emitters and photodetectors for optical communication based on two-dimensional materials. *Opt. Mater. Express 11*, 3298–3320 (2021)
227. You, J. et al. Hybrid/integrated silicon photonics based on 2D materials in optical communication nanosystems. *Laser Photon. Rev.* 14, 2000239 (2020)
228. Fan, X. et al. Broken symmetry induced strong nonlinear optical effects in spiral WS_2 nanosheets. *ACS Nano 11*, 4892–4898 (2017)
229. Wang, B. et al. High-efficiency second-harmonic and sum-frequency generation in a silicon nitride microring integrated with few-layer GaSe. *ACS Photon 9*, 1671–1678 (2022)
230. Xiang, C. et al. Narrow-linewidth III–V/Si/Si_3N_4 laser using multilayer heterogeneous integration. *Optica 7*, 20–21 (2020)
231. Xuan, Y. et al. High-Q silicon nitride microresonators exhibiting low-power frequency comb initiation. *Optica 3*, 1171–1180 (2016)
232. Ji, X. et al. Ultra-low-loss on-chip resonators with sub-milliwatt parametric oscillation threshold. *Optica 4*, 619–624 (2017)
233. Liu, J. et al. Ultralow-power chip-based soliton microcombs for photonic integration. *Optica 5*, 1347–1353 (2018)
234. Ye, Z., Twayana, K., Andrekson, P. A. & Torres-Company, V. High-Q Si_3N_4 microresonators based on a subtractive processing for Kerr nonlinear optics. *Opt. Express 27*, 35719–35727 (2019)
235. Dirani, H. E. et al. Ultralow-loss tightly confining Si_3N_4 waveguides and high-Q microresonators. *Opt. Express 27*, 30726–30740 (2019)
236. Li, Q., Davanço, M. & Srinivasan, K. Efficient and low-noise single-photon-level frequency conversion interfaces using silicon nanophotonics. *Nat. Photonics 10*, 406–414 (2016)
237. Lu, X. et al. Efficient telecom-to-visible spectral translation through ultralow power nonlinear nanophotonics. *Nat. Photonics 13*, 593–601 (2019)
238. Hausmann, B. J. M., Bulu, I., Venkataraman, V., Deotare, P. & Lončar, M. Diamond nonlinear photonics. *Nat. Photonics 8*, 369–374 (2014)
239. Jung, H. et al. Tantala Kerr-nonlinear integrated photonics. Preprint at http://arxiv.org/abs/2007.12958 (2020)
240. Guidry, M. A. et al. Optical parametric oscillation in silicon carbide nanophotonics. *Optica 7*, 1139–1142 (2020)
241. Pu, M., Ottaviano, L., Semenova, E. & Yvind, K. Efficient frequency comb generation in AlGaAs-on-insulator. *Optica 3*, 823–826 (2016)
242. Chang, L. et al. Ultra-efficient frequency comb generation in AlGaAs-on-insulator microresonators. *Nat. Commun. 11*, 1331 (2020)
243. Wilson, D. J. et al. Integrated gallium phosphide nonlinear photonics. *Nat. Photon.* 14, 57–62 (2020)
244. Fang, Z. et al. Efficient electro-optical tuning of an optical frequency microcomb on a monolithically integrated high-Q lithium niobate microdisk. *Opt. Lett. 44*, 5953–5956 (2019)
245. Jung, H., Fong, K. Y., Xiong, C. & Tang, H. X. Electrical tuning and switching of an optical frequency comb generated in aluminum nitride microring resonators. *Opt. Lett. 39*, 84–87 (2014)
246. Guo, X. et al. Second-harmonic generation in aluminum nitride microrings with 2500%/W conversion efficiency. *Optica 3*, 1126–1131 (2016)
247. Liu, X. et al. Integrated high-Q crystalline AlN microresonators for broadband Kerr and Raman frequency combs. *ACS Photonics 5*, 1943–1950 (2018)

248. Moss, D. J., Morandotti, R., Gaeta, A. L. & Lipson, M. New CMOS-compatible platforms based on silicon nitride and Hydex for nonlinear optics. *Nat. Photon. 7*, 597–607 (2013)
249. Brasch, V., Chen, Q.F., Schiller, S. & Kippenberg, T.J. Radiation hardness of high-Q silicon nitride microresonators for space compatible integrated optics. *Opt. Express 22*, 30786–30794 (2014)
250. Gyger, F. et al. Observation of stimulated Brillouin scattering in silicon nitride integrated waveguides. *Phys. Rev. Lett. 124*, 013902 (2020)
251. Brasch, V. et al. Photonic chip–based optical frequency comb using soliton Cherenkov radiation. *Science 351*, 357–360 (2016)
252. Kovach, A. et al. Emerging material systems for integrated optical Kerr frequency combs. *Adv. Opt. Photonics 12*, 135–222 (2020)
253. Spencer, D. T., Bauters, J. F., Heck, M. J. R. & Bowers, J. E. Integrated waveguide coupled Si_3N_4 resonators in the ultrahigh-Q regime. *Optica 1*, 153–157 (2014)
254. Luke, K., Okawachi, Y., Lamont, M. R. E., Gaeta, A. L. & Lipson, M. Broadband mid-infrared frequency comb generation in a Si_3N_4 microresonator. *Opt. Lett. 40*, 4823–4826 (2015)
255. Bauters, J. F. et al. Planar waveguides with less than 0.1 dB/m propagation loss fabricated with wafer bonding. *Opt. Express 19*, 24090–24101 (2011)
256. Wójtewicz, S. et al. Response of an optical cavity to phase-controlled incomplete power switching of nearly resonant incident light. *Opt. Express 26*, 5644–5654 (2018)
257. Misra, J., Saha, I.: Artificial neural networks in hardware: a survey of two decades of progress. Neurocomputing *74*(1), 239255 (2010)
258. Intel delivers Real Time AI in Microsoft's accelerated deep learning platform. [Online]. Available: https://newsroom.intel.com/news/inteldelivers-real-time-aimicrosofts-accelerated-deep-learning-platform/
259. Rajendran, B., et al.: Low-power neuromorphic hardware for signal processing applications: A review of architectural and system-level design approaches. IEEE Signal Process. Mag. *36*(6), 97–110 (2019)
260. Yao, P., Wu, H., Gao, B., Tang, J., Zhang, Q., Zhang, W.: Fully hardware implemented memristor convolutional neural network. Nature *577*, 641–646 (2020)
261. Feng, C., Gu, J., Zhu, H., Ying, Z., Zhao, Z., Pan, D.Z., Chen, R.T.: Silicon photonic subspace neural chip for hardware-efficient deep learning. arXiv preprint arXiv:2111.06705. (2021)
262. Sunny, F.P., et al.: A survey on silicon photonics for deep learning. ACM J. Emerg. Technol. Comput. Syst. *17*, 1–57 (2021)
263. Mubarak Ali, M., Madhupriya, G., Indhumathi, R., Krishnamoorthy, P, Photonic processing core for reconfigurable electronic–photonic integrated circuit. In: Arunachalam, V., Sivasankaran, K. (eds.) Microelectronic Devices, Circuits and Systems. ICMDCS 2021. Communications in Computer and Information Science, vol. 1392. Springer, Singapore (2021)
264. Meerasha, M.A., Ganesh, M., Pandiyan, K.: Reconfigurable quantum photonic convolutional neural network layer utilizing photonic gate and teleportation mechanism. Opt. Quant. Electron. *54*, 770 (2022). https://doi.org/10.1007/s11082-022-04168-8
265. Ali, M.M., Madhupriya, G., Indhumathi, R., Krishnamoorthy, P.: Performance enhancement of 8×8 dilated banyan network using crosstalk suppressed GMZI crossbar photonic switches. Photonic Netw. Commun. *42*, 123–133 (2021)
266. Rogers, C. et al. A universal 3d imaging sensor on a silicon photonics platform. *Nature 590*, 256–261 (2021)
267. Zhang, G. et al. Demonstration of high output power DBR laser integrated with SOA for the FMCW LiDAR system. *Opt. Express 30*, 2599–2609 (2022)
268. Poulton, C. V. et al. Long-range lidar and free-space data communication with high-performance optical phased arrays. *IEEE J. Sel. Top. Quantum Electron. 25*, 1–8 (2019)

269. Li, M. et al. Integrated Pockels laser. *Nat. Commun.* 13, 5344 (2022)
270. Snigirev, V. et al. Ultrafast tunable lasers using lithium niobate integrated photonics. *Nature* 615, 411–417 (2023)
271. Zhang, X., Pouls, J. & Wu, M. C. Laser frequency sweep linearization by iterative learning pre-distortion for FMCW lidar. *Opt. Express* 27, 9965–9974 (2019)
272. Carroll, L. et al. Photonic packaging: transforming silicon photonic integrated circuits into photonic devices. *Appl. Sci.* 6, 426 (2016)
273. Xiang, C. et al. High-performance silicon photonics using heterogeneous integration. *IEEE J. Sel. Top. Quantum Electron.* 28, 1–15 (2022)
274. Liang, D. & Bowers, J. E. Recent progress in heterogeneous III–V-on-silicon photonic integration. *Light Adv. Manuf.* 2, 59–83 (2021)
275. Marshall, O. et al. Heterogeneous integration on silicon photonics. *Proc. IEEE* 106, 2258–2269 (2018)
276. Pintus, P. et al. Microring-based optical isolator and circulator with integrated electromagnet for silicon photonics. *J. Lightw. Technol.* 35, 1429–1437 (2017)
277. Shulaker, M. M. et al. Three-dimensional integration of nanotechnologies for computing and data storage on a single chip. *Nature* 547, 74–78 (2017)
278. Rachmady, W. et al. 300 mm heterogeneous 3D integration of record performance layer transfer germanium PMOS with silicon NMOS for low power high performance logic applications. In *2019 IEEE International Electron Devices Meeting* 29.7.1–29.7.4 (IEEE, 2019)
279. Sacher, W. D. et al. Monolithically integrated multilayer silicon nitride-on-silicon waveguide platforms for 3-D photonic circuits and devices. *Proc. IEEE* 106, 2232–2245 (2018)
280. Xiang, C. et al. Laser soliton microcombs heterogeneously integrated on silicon. *Science* 373, 99–103 (2021)
281. Zhang, Z. et al. High-speed coherent optical communication with isolator-free heterogeneous Si/III–V lasers. *J. Lightw. Technol.* 38, 6584–6590 (2020)
282. Gomez, S. et al. High coherence collapse of a hybrid III–V/Si semiconductor laser with a large quality factor. *J. Phys. Photon.* 2, 025005 (2020)
283. Guo, J. et al. Chip-based laser with 1-hertz integrated linewidth. *Sci. Adv.* 8, eabp9006 (2022)
284. Jin, N. et al. Micro-fabricated mirrors with finesse exceeding one million. *Optica* 9, 965–970 (2022)
285. Hulme, J. et al. Fully integrated microwave frequency synthesizer on heterogeneous silicon–III/V. *Opt. Express* 25, 2422–2431 (2017)
286. Kittlaus, E. A. et al. A low-noise photonic heterodyne synthesizer and its application to millimeter-wave radar. *Nat. Commun.* 12, 4397 (2021)
287. Gundavarapu, S. et al. Sub-hertz fundamental linewidth photonic integrated Brillouin laser. *Nat. Photonics* 13, 60–67 (2019)
288. Liu, Y. et al. A photonic integrated circuit-based erbium-doped amplifier. *Science* 376, 1309–1313 (2022)
289. Liang, W. et al. Resonant microphotonic gyroscope. *Optica* 4, 114–117 (2017)
290. Spencer, D. T. et al. An optical-frequency synthesizer using integrated photonics. *Nature* 557, 81–85 (2018)
291. Ramadan, T. A. & Osgood, R. M. Adiabatic couplers: design rules and optimization. *J. Lightw. Technol.* 16, 277 (1998)
292. Bogaerts, W. et al. Programmable photonic circuits. *Nature* 586, 207–216 (2020)
293. Pérez, D., Gasulla, I. & Capmany, J. Programmable multifunctional integrated nanophotonics. *Nanophotonics* 7, 1351–1371 (2018)
294. Miller, D. A. B. Self-configuring universal linear optical component [invited]. *Photon. Res.* 1, 1–15 (2013)

295. Edinger, P. et al. Silicon photonic microelectromechanical phase shifters for scalable programmable photonics. *Opt. Lett. 46*, 5671–5674 (2021)
296. Henriksson, J. et al. Digital silicon photonic MEMS phase-shifter. In *2018 International Conference on Optical MEMS and Nanophotonics (OMN)* 1–2 (IEEE, 2018)
297. Sattari, H. et al. Silicon photonic MEMS phase-shifter. *Opt. Express 27*, 18959–18969 (2019)
298. Schuck, C., Grottke, T., Hartmann, W. & Pernice, W. H. P. Optoelectromechanical phase shifter with low insertion loss and a 13π tuning range. *Opt. Express 29*, 5525–5537 (2021)
299. Ramey, C. et al. Dual slot-mode NOEM phase shifter. *Opt. Express 29*, 19113–19119 (2021)
300. Quack, N. et al. Integrated silicon photonic MEMS. *Microsyst. Nanoeng. 9*, 27 (2023)
301. Edinger, P. et al. Vacuum-sealed silicon photonic MEMS tunable ring resonator with an independent control over coupling and phase. *Opt. Express 31*, 6540–6551 (2023)
302. Pérez-López, D., Gutierrez, A. M., Sánchez, E., DasMahapatra, P. & Capmany, J. Integrated photonic tunable basic units using dual-drive directional couplers. *Opt. Express 27*, 38071–38086 (2019)
303. Pérez, D. et al. Multipurpose silicon photonics signal processor core. *Nat. Commun. 8*, 636 (2017)
304. Yap, K. P. et al. Correlation of scattering loss, sidewall roughness and waveguide width in silicon-on-insulator (SOI) ridge waveguides. *J. Lightwave Technol. 27*, 3999–4008 (2009)
305. Riemensberger, Johann; Kuznetsov, Nikolai; Liu, Junqiu; He, Jijun; Wang, Rui Ning; Kippenberg, Tobias J. A photonic integrated continuous-travelling-wave parametric amplifier. Nature (2022). https://doi.org/10.1038/s41586-022-05329-1
306. Sharma, Sumit; Roy, Sudip A survey on design and synthesis techniques for photonic integrated circuits. The Journal of Supercomputing (2020). https://doi.org/10.1007/s11227-020-03430-8
307. Sharif Azadeh, Saeed; Mak, Jason C. C.; Chen, Hong; Luo, Xianshu; Chen, Fu-Der; Chua, Hongyao; Weiss, Frank; Alexiev, Christopher; Stalmashonak, Andrei; Jung, Youngho; Straguzzi, John N.; Lo, Guo-Qiang; Sacher, Wesley D.; Poon, Joyce K. S. Microcantilever-integrated photonic circuits for broadband laser beam scanning. Nature Communications (2023). https://doi.org/10.1038/s41467-023-38260-8
308. Maiti, R.; Patil, C.; Saadi, M. A. S. R.; Xie, T.; Azadani, J. G.; Uluutku, B.; Amin, R.; Briggs, A. F.; Miscuglio, M.; Van Thourhout, D.; Solares, S. D.; Low, T.; Agarwal, R.; Bank, S. R.; Sorger, V. J. Strain-engineered high-responsivity MoTe$_2$ photodetector for silicon photonic integrated circuits. Nature Photonics (2020). https://doi.org/10.1038/s41566-020-0647-4
309. Liu, Junqiu; Lucas, Erwan; Raja, Arslan S.; He, Jijun; Riemensberger, Johann; Wang, Rui Ning; Karpov, Maxim; Guo, Hairun; Bouchand, Romain; Kippenberg, Tobias J. Photonic microwave generation in the X- and K-band using integrated soliton microcombs. Nature Photonics (2020). https://doi.org/10.1038/s41566-020-0617-x
310. Chen, Minjia; Wang, Yizhi; Yao, Chunhui; Wonfor, Adrian; Yang, Shuai; Penty, Richard; Cheng, Qixiang I/O-efficient iterative matrix inversion with photonic integrated circuits. Nature Communications (2024). https://doi.org/10.1038/s41467-024-50302-3
311. Lee, Jongmin; Ding, Roger; Christensen, Justin; Rosenthal, Randy R.; Ison, Aaron; Gillund, Daniel P.; Bossert, David; Fuerschbach, Kyle H.; Kindel, William; Finnegan, Patrick S.; Wendt, Joel R.; Gehl, Michael; Kodigala, Ashok; McGuinness, Hayden; Walker, Charles A.; Kemme, Shanalyn A.; Lentine, Anthony; Biedermann, Grant; Schwindt, Peter D. D. A compact cold-atom interferometer with a high data-rate grating magneto-optical trap and a photonic-integrated-circuit-compatible laser system. Nature Communications (2022). https://doi.org/10.1038/s41467-022-31410-4
312. Rank, Elisabet A.; Nevlacsil, Stefan; Muellner, Paul; Hainberger, Rainer; Salas, Matthias; Gloor, Stefan; Duelk, Marcus; Sagmeister, Martin; Kraft, Jochen; Leitgeb, Rainer A.; Drexler,

Wolfgang In vivo human retinal swept source optical coherence tomography and angiography at 830 nm with a CMOS compatible photonic integrated circuit. Scientific Reports (2021). https://doi.org/10.1038/s41598-021-00637-4
313. Dong, Mark; Clark, Genevieve; Leenheer, Andrew J.; Zimmermann, Matthew; Dominguez, Daniel; Menssen, Adrian J.; Heim, David; Gilbert, Gerald; Englund, Dirk; Eichenfield, Matt High-speed programmable photonic circuits in a cryogenically compatible, visible–near-infrared 200 mm CMOS architecture. Nature Photonics (2021). https://doi.org/10.1038/s41566-021-00903-x
314. Wan, Noel H.; Lu, Tsung-Ju; Chen, Kevin C.; Walsh, Michael P.; Trusheim, Matthew E.; De Santis, Lorenzo; Bersin, Eric A.; Harris, Isaac B.; Mouradian, Sara L.; Christen, Ian R.; Bielejec, Edward S.; Englund, Dirk Large-scale integration of artificial atoms in hybrid photonic circuits. Nature (2020). https://doi.org/10.1038/s41586-020-2441-3
315. Hogle, C. W.; Dominguez, D.; Dong, M.; Leenheer, A.; McGuinness, H. J.; Ruzic, B. P.; Eichenfield, M.; Stick, D. High-fidelity trapped-ion qubit operations with scalable photonic modulators. npj Quantum Information (2023). https://doi.org/10.1038/s41534-023-00737-1
316. Chen, Rui; Fang, Zhuoran; Perez, Christopher; Miller, Forrest; Kumari, Khushboo; Saxena, Abhi; Zheng, Jiajiu; Geiger, Sarah J.; Goodson, Kenneth E.; Majumdar, Arka Non-volatile electrically programmable integrated photonics with a 5-bit operation. Nature Communications (2023). https://doi.org/10.1038/s41467-023-39180-3
317. Jooken, Stijn; Zinoviev, Kirill; Yurtsever, Günay; De Proft, Anabel; de Wijs, Koen; Jafari, Zeinab; Lebanov, Ana; Jeevanandam, Gaudhaman; Kotyrba, Mateusz; Gorjup, Erwin; Fondu, Jelle; Lagae, Liesbet; Libbrecht, Sarah; Van Dorpe, Pol; Verellen, Niels On-chip flow cytometer using integrated photonics for the detection of human leukocytes. Scientific Reports (2024). https://doi.org/10.1038/s41598-024-60708-0
318. Rank, Elisabet A.; Sentosa, Ryan; Harper, Danielle J.; Salas, Matthias; Gaugutz, Anna; Seyringer, Dana; Nevlacsil, Stefan; Maese-Novo, Alejandro; Eggeling, Moritz; Muellner, Paul; Hainberger, Rainer; Sagmeister, Martin; Kraft, Jochen; Leitgeb, Rainer A.; Drexler, Wolfgang Toward optical coherence tomography on a chip: in vivo three-dimensional human retinal imaging using photonic integrated circuit-based arrayed waveguide gratings. Light: Science & Applications (2021). https://doi.org/10.1038/s41377-020-00450-0
319. Wang, Chengli; Li, Zihan; Riemensberger, Johann; Lihachev, Grigory; Churaev, Mikhail; Kao, Wil; Ji, Xinru; Zhang, Junyin; Blesin, Terence; Davydova, Alisa; Chen, Yang; Huang, Kai; Wang, Xi; Ou, Xin; Kippenberg, Tobias J. Lithium tantalate photonic integrated circuits for volume manufacturing. Nature (2024). https://doi.org/10.1038/s41586-024-07369-1
320. Snigirev, Viacheslav; Riedhauser, Annina; Lihachev, Grigory; Churaev, Mikhail; Riemensberger, Johann; Wang, Rui Ning; Siddharth, Anat; Huang, Guanhao; Möhl, Charles; Popoff, Youri; Drechsler, Ute; Caimi, Daniele; Hönl, Simon; Liu, Junqiu; Seidler, Paul; Kippenberg, Tobias J. Ultrafast tunable lasers using lithium niobate integrated photonics. Nature (2023). https://doi.org/10.1038/s41586-023-05724-2
321. Xiang, Chao; Guo, Joel; Jin, Warren; Wu, Lue; Peters, Jonathan; Xie, Weiqiang; Chang, Lin; Shen, Boqiang; Wang, Heming; Yang, Qi-Fan; Kinghorn, David; Paniccia, Mario; Vahala, Kerry J.; Morton, Paul A.; Bowers, John E. High-performance lasers for fully integrated silicon nitride photonics. Nature Communications (2021). https://doi.org/10.1038/s41467-021-26804-9
322. Lin, Chupao; Peñaranda, Juan Santo Domingo; Dendooven, Jolien; Detavernier, Christophe; Schaubroeck, David; Boon, Nico; Baets, Roel; Le Thomas, Nicolas UV photonic integrated circuits for far-field structured illumination autofluorescence microscopy. Nature Communications (2022). https://doi.org/10.1038/s41467-022-31989-8

323. Zhu, Cheng-Yi; Zhang, Zimeng; Qin, Jing-Kai; Wang, Zi; Wang, Cong; Miao, Peng; Liu, Yingjie; Huang, Pei-Yu; Zhang, Yao; Xu, Ke; Zhen, Liang; Chai, Yang; Xu, Cheng-Yan Two-dimensional semiconducting SnP_2Se_6 with giant second-harmonic-generation for monolithic on-chip electronic-photonic integration. Nature Communications (2023). https://doi.org/10.1038/s41467-023-38131-2
324. Liu, Junqiu; Huang, Guanhao; Wang, Rui Ning; He, Jijun; Raja, Arslan S.; Liu, Tianyi; Engelsen, Nils J.; Kippenberg, Tobias J. High-yield, wafer-scale fabrication of ultralow-loss, dispersion-engineered silicon nitride photonic circuits. Nature Communications (2021). https://doi.org/10.1038/s41467-021-21973-z
325. Mosses, A.; Joe Prathap, P. M. Analysis and codesign of electronic–photonic integrated circuit hardware accelerator for machine learning application. Journal of Computational Electronics (2024). https://doi.org/10.1007/s10825-023-02123-8
326. Lukashchuk, Anton; Yildirim, Halil Kerim; Bancora, Andrea; Lihachev, Grigory; Liu, Yang; Qiu, Zheru; Ji, Xinru; Voloshin, Andrey; Bhave, Sunil A.; Charbon, Edoardo; Kippenberg, Tobias J. Photonic-electronic integrated circuit-based coherent LiDAR engine. Nature Communications (2024). https://doi.org/10.1038/s41467-024-47478-z
327. Xiang, Chao; Jin, Warren; Terra, Osama; Dong, Bozhang; Wang, Heming; Wu, Lue; Guo, Joel; Morin, Theodore J.; Hughes, Eamonn; Peters, Jonathan; Ji, Qing-Xin; Feshali, Avi; Paniccia, Mario; Vahala, Kerry J.; Bowers, John E. 3D integration enables ultralow-noise isolator-free lasers in silicon photonics. Nature (2023). https://doi.org/10.1038/s41586-023-06251-w
328. Kim, Dong Uk; Park, Young Jae; Kim, Do Yun; Jeong, Youngjae; Lim, Min Gi; Hong, Myung Seok; Her, Man Jae; Rah, Yoonhyuk; Choi, Dong Ju; Han, Sangyoon; Yu, Kyoungsik Programmable photonic arrays based on microelectromechanical elements with femtowatt-level standby power consumption. Nature Photonics (2023). https://doi.org/10.1038/s41566-023-01327-5
329. Xue, C.-X. et al. 15.4 A 22 nm 2 Mb ReRAM compute-in-memory macro with 121-28TOPS/W for multibit MAC computing for tiny AI edge devices. In *2020 IEEE International Solid-State Circuits Conference (ISSCC) Digest of Technical Papers* 244–245 (IEEE, 2020)
330. Xue, C.-X. et al. 24.1 A 1 Mb multibit ReRAM computing-in-memory macro with 14.6 ns parallel MAC computing time for CNN based AI edge processors. In *2019 IEEE International Solid-State Circuits Conference (ISSCC) Digest of Technical Papers* 388–390 (IEEE, 2019)
331. Chen, W.-H. et al. CMOS-integrated memristive non-volatile computing-in-memory for AI edge processors. *Nat. Electron.* 2, 420–428 (2019)
332. Tang, K.-T. et al. Considerations of integrating computing-in-memory and processing-in-sensor into convolutional neural network accelerators for low-power edge devices. In *2019 IEEE Symposium on VLSI Technology* T166–T167 (IEEE, 2019)
333. Mochida, R. et al. A 4M synapses integrated analog ReRAM based 66.5 TOPS/W neural-network processor with cell current controlled writing and flexible network architecture. In *2018 IEEE Symposium on VLSI Technology* 175–176 (IEEE, 2018)
334. Deng, L. The MNIST database of handwritten digit images for machine learning research. *IEEE Signal Process. Mag.* 29, 141–142 (2012)
335. Krizhevsky, A. *Learning Multiple Layers of Features from Tiny Images* (Univ. Toronto, 2009); http://www.cs.toronto.edu/~kriz/learning-features-2009-TR.pdf
336. Wan, W. et al. 33.1 A 74 TMACS/W CMOS-RRAM neurosynaptic core with dynamically reconfigurable dataflow and in-situ transposable weights for probabilistic graphical models. In *2020 IEEE International Solid-State Circuits Conference (ISSCC) Digest of Technical Papers* 498–499 (IEEE, 2020)

337. Liu, Q. et al. 33.2 A fully integrated analog ReRAM based 78.4TOPS/W compute-in-memory chip with fully parallel MAC computing. In *2020 IEEE International Solid-State Circuits Conference (ISSCC) Digest of Technical Papers* 500–502 (IEEE, 2020)
338. Cai, F. et al. A fully integrated reprogrammable memristor–CMOS system for efficient multiply–accumulate operations. *Nat. Electron.* 2, 290–299 (2019)
339. Li, C. et al. Analogue signal and image processing with large memristor crossbars. *Nat. Electron.* 1, 52–59 (2018)
340. Wang, Z. et al. Fully memristive neural networks for pattern classification with unsupervised learning. *Nat. Electron.* 1, 137–145 (2018)
341. Ambrogio, S. et al. Equivalent-accuracy accelerated neural-network training using analogue memory. *Nature 558*, 60–67 (2018)
342. Wu, F. et al. Brain-inspired computing exploiting carbon nanotube FETs and resistive RAM: hyperdimensional computing case study. In *2018 IEEE International Solid-State Circuits Conference (ISSCC) Digest of Technical Papers* 492–494 (IEEE, 2018)
343. Zidan, M.-A. et al. The future of electronics based on memristive systems. *Nat. Electron.* 1, 22–29 (2018)
344. Ielmini, D. et al. In-memory computing with resistive switching devices. *Nat. Electron.* 1, 333–343 (2018)
345. Yao, P. et al. Face classification using electronic synapses. *Nat. Commun.* 8, 15199 (2017)
346. Sheridan, P. et al. Sparse coding with memristor networks. *Nat. Nanotechnol.* 12, 784–789 (2017)
347. Li, H. et al. Hyperdimensional computing with 3D VRRAM in-memory kernels: device-architecture co-design for energy-efficient, error-resilient language recognition. In *2016 IEEE International Electron Devices Meeting (IEDM)* 16.1.1–16.1.4 (IEEE, 2016)
348. Chen, B. et al. Efficient in-memory computing architecture based on crossbar arrays. In *2015 IEEE International Electron Devices Meeting (IEDM)* 17.5.1–17.5.4 (IEEE, 2015)
349. Prezioso, M. et al. Training and operation of an integrated neuromorphic network based on metal-oxide memristors. *Nature 521*, 61–64 (2015)
350. Wong, H.-S. P. et al. Memory leads the way to better computing. *Nat. Nanotechnol.* 10, 191–194 (2015)
351. Yang, J. J. et al. Memristive devices for computing. *Nat. Nanotechnol.* 8, 13–24 (2013)
352. Borghetti, J. et al. 'Memristive' switches enable 'stateful' logic operations via material implication. *Nature 464*, 873–876 (2010)
353. Ney, A. et al. Programmable computing with a single magnetoresistive element. *Nature 425*, 485–487 (2003)
354. Chou, C.-C. et al. A 22 nm 96KX144 RRAM macro with a self tracking reference and a low ripple charge pump to achieve a configurable read window and a wide operating voltage range. In *2020 IEEE Symposium on VLSI Circuits* 1–2 (IEEE, 2020)
355. Dong, Q. et al. 15.3 A 351TOPS/W and 372.4GOPS compute-in-memory SRAM macro in 7 nm FinFET CMOS for machine-learning applications. In *2020 IEEE International Solid-State Circuits Conference (ISSCC) Digest of Technical Papers* 242–244 (IEEE, 2020)
356. Gonugondla, S. K. et al. A 42pJ/decision 3.12TOPS/W robust in-memory machine learning classifier with on-chip training. In *2018 IEEE International Solid-State Circuits Conference (ISSCC) Digest of Technical Papers* 490–492 (IEEE, 2018)
357. Biswas, A. et al. Conv-RAM: an energy-efficient SRAM with embedded convolution computation for low-power CNN-based machine learning applications. In *2018 IEEE International Solid-State Circuits Conference (ISSCC) Digest of Technical Papers* 488–490 (IEEE, 2018)
358. Levisse, A. et al. Write termination circuits for RRAM: a holistic approach from technology to application considerations. *IEEE Access 8*, 109297–109308 (2020)

359. Chang, M.-F. et al. 19.4 Embedded 1 Mb ReRAM in 28 nm CMOS with 0.27-to-1 V read using swing-sample-and-couple sense amplifier and self-boost-write-termination scheme. In *2014 IEEE International Solid-State Circuits Conference (ISSCC) Digest of Technical Papers* 332–333 (IEEE, 2014)
360. Liu, Y. et al. 4.7 A 65 nm ReRAM-enabled nonvolatile processor with 6× reduction in restore time and 4× higher clock frequency using adaptive data retention and self-write-termination nonvolatile logic. In *2016 IEEE International Solid-State Circuits Conference (ISSCC) Digest of Technical Papers* 84–86 (IEEE, 2016)
361. Wu, T. F. et al. 14.3 A 43pJ/cycle non-volatile microcontroller with 4.7 µs shutdown/wake-up integrating 2.3-bit/cell resistive RAM and resilience techniques. In *2019 IEEE International Solid-State Circuits Conference (ISSCC) Digest of Technical Papers* 226–228 (IEEE, 2019)
362. Jain, P. et al. 13.2 A 3.6 Mb 10.1 Mb/mm^2 embedded non-volatile ReRAM macro in 22 nm FinFET technology with adaptive forming/set/reset schemes yielding down to 0.5 V with sensing time of 5 ns at 0.7 V. In *2019 IEEE International Solid-State Circuits Conference (ISSCC) Digest of Technical Papers* 212–214 (IEEE, 2019)
363. Lee, C.-F. et al. A 1.4 Mb 40-nm embedded ReRAM macro with 0.07 um^2 bit cell, 2.7 mA/100 MHz low-power read and hybrid write verify for high endurance application. In *2017 IEEE Asian Solid-State Circuits Conference (A-SSCC)* 9–12 (IEEE, 2017)
364. Seo, S. et al. Artificial optic-neural synapse for colored and color-mixed pattern recognition. *Nat. Commun. 9*, 5106 (2018)
365. Zhou, F. et al. Optoelectronic resistive random access memory for neuromorphic vision sensors. *Nat. Nanotechnol. 14*, 776–782 (2019)
366. Yu, S. et al. Binary neural network with 16 Mb RRAM macro chip for classification and online training. In *2016 IEEE International Electron Devices Meeting (IEDM)* 16.2.1–16.2.4 (IEEE, 2016)
367. Cassinerio, M. et al. Logic computation in phase change materials by threshold and memory switching. *Adv. Mater. 25*, 5975–5980 (2013)
368. Abbey, T. et al. An embedded environmental control micro-chamber system for RRAM memristor characterisation. In *2018 IEEE International Symposium on Circuits and Systems (ISCAS)* 1–4 (IEEE, 2018)
369. Boybat, I. et al. Neuromorphic computing with multi-memristive synapses. *Nat. Commun. 9*, 2514 (2018)
370. Gallo, M. L. et al. Mixed-precision in-memory computing. *Nat. Electron. 1*, 246–253 (2018)
371. Xue, C.-X. et al. A CMOS-integrated compute-in-memory macro based on resistive random-access memory for AI edge devices. *Nat. Electron 4*, 81–90 (2021)
372. Chen, W.-H. et al. A 65 nm 1 Mb nonvolatile computing-in-memory ReRAM macro with sub-16ns multiply-and-accumulate for binary DNN AI edge processors. In *2018 IEEE International Solid-State Circuits Conference (ISSCC) Digest of Technical Papers* 494–496 (IEEE, 2018)
373. Chatterjee A, Polgreen T (1991) A low-voltage triggering SCR for on-chip ESD protection at output and input pads. IEEE Electron Device Lett 12:21–22. https://doi.org/10.1109/55.75685
374. Ker MD, Wang KF, Joe MC, Chu YH, Wu TS (1995) Area-efficient CMOS output buffer with enhanced high ESD reliability for deep submicron CMOS ASIC. In: Proceedings of 8th IEEE international ASIC conference and exhibition, pp 123–126
375. Ker M-D, Wu C-Y, Chang H-H (1996) Complementary-LVTSCR ESD protection circuit for submicron CMOS VLSI/ULSI. IEEE Trans Electron Devices 43:588–598
376. Notermans G, Kuper F, Luchis JM (1997) Using an SCR as ESD protection without latch-up danger. Microelectron Reliab 37:1457–1460

377. Wu D-J, Miao M, Zeng J, Han Y, Liou JJ (2012) High-holding voltage silicon-controlled rectifier for ESD applications. IEEE Electron Device Lett 33(10):1345–1347. https://doi.org/10.1109/LED.2012.2208934
378. Huang C-Y, Kao T-C, Lee J-H et al (2014) Simple scheme to increase hold voltage for silicon-controlled rectifier. IEEE Electron Lett 50(3):200–202. https://doi.org/10.1049/el.2013.1853
379. Sleeter DJ, Enlow EW (1992) The relationship of holding points and a general solution for CMOS latchup. IEEE Trans Electron Devices 39(11):2592–2599. https://doi.org/10.1109/16.163468
380. Lee JH, Weng WT, Shih JR, Yu KF, Ong TC (2004) The positive trigger voltage lowering effect for latch-up. In: International symposium on the physical and failure analysis of integrated circuits (IPFA). https://doi.org/10.1109/IPFA.2004.1345550
381. Gavaskar, K., Ragupathy, U. S., & Malini, V. (2019). Proposed design of 1 KB memory array structure for cache memories. *Wireless Personal Communications, 109*(2), 823–847
382. Nanda, U., Acharya, D. P., Rout, P. K., Nayak, D., & Jena, B. (2020). Performance-linked phase-locked loop architectures: recent developments. *Advanced VLSI Design and Testability Issues*, pp. 271–290
383. Sharma, J., & Krishnaswamy, H. (2019). A 2.4-GHz reference-sampling phase-locked loop that simultaneously achieves low-noise and low-spur performance. *IEEE Journal of Solid-State Circuits, 54*(5), 1407–1424
384. Malathi, D., & Gomathi, M. (2019). Design of inductively degenerated common source RF CMOS low noise amplifier. *Sādhanā, 44*(1), 1–9
385. Devi, T. K., Priyanka, E. B., Sakthivel, P., & Sagayaraj, A. S. (2021). Sleepy keeper style based low power VLSI architecture of a viterbi decoder applying for the wireless LAN Operation sustainability. *Analog Integrated Circuits and Signal Processing*, 1–13
386. Metange, P. N., & Khanchandani, K. B. (2019). Ultra-low power hybrid PLL frequency synthesizer with lock check provisioning efficient phase noise. *Journal of Information Science & Engineering, 35*(6)
387. Berber, Z., Kameche, S., & Benkhelifa, E. (2019). High tolerance of charge pump leakage current in Integer-N PLL frequency synthesizer for 5G networks. *Simulation Modelling Practice and Theory, 95*, 134–147
388. Zhao, H., & Mandal, S. (2019). A fast-settling integer-N frequency synthesizer using switched-gain control. *IEEE Transactions on Circuits and Systems I: Regular Papers, 67*(4), 1344–1357
389. Markulic, N., Raczkowski, K., Craninckx, J., & Wambacq, P. (2019). *Digital subsampling phase lock techniques for frequency synthesis and polar transmission*. Springer
390. Priyanka, E. B., Thangavel, S., & Pratheep, V. G. (2020). Enhanced digital synthesized phase locked loop with high frequency compensation and clock generation. *Sensing and Imaging, 21*(1), 1–12
391. Ko, H. G., Bae, W., Jeong, G. S., & Jeong, D. K. (2019). Reference spur reduction techniques for a phase-locked loop. *IEEE Access, 7*, 38035–38043
392. Casson, A.J.; Abdullal, M.; Dulabh, M.; Kohli, S.; Krachunov, S.; Trimble, E.: Electroencephalogram. In: Tamura, T., Chen, W. (eds.) Seamless Healthcare Monitoring, Chap. 2, pp. 45–81. Springer, Berlin (2018)
393. Bronzino, J.D. (ed.): Principles of electroencephalography. In: The Biomedical Engineering Handbook, Chap. 3, vol. 1, 2nd edn. CRC and IEEE Press, USA (2000)
394. Feng, J.; Yan, N.; Min, H.: A low power low noise amplifier for EEG/ECG signal recording applications. In: IEEE International Conference on ASIC, pp. 145–148, Xiamen, China (2011)
395. Yin, M.; Ghovanloo, M.: A low noise preamplifier with adjustable gain and bandwidth for biopotential recording applications. In: IEEE International Symposium on Circuits and Systems, pp. 321–324, New Orleans, LA (2007)

396. Razavi, B.: Design of Analog CMOS Integrated Circuits. McGraw Hill, New York (2001)
397. Harrison, R.R.; Charles, C.: A low-power low-noise CMOS amplifier for neural recording applications. IEEE J. Solid-State Circuits *38*(6), 958–965 (2003)
398. Lucas, A. Ising formulations of many NP problems. *Front. Phys. 2*, 5 (2014)
399. Mohseni, N., McMahon, P. L. & Byrnes, T. Ising machines as hardware solvers of combinatorial optimization problems. *Nat. Rev. Phys. 4*, 363–379 (2022)
400. Ueyoshi, K., Marukame, T., Asai, T., Motomura, M. & Schmid, A. FPGA implementation of a scalable and highly parallel architecture for restricted Boltzmann machines. *Circuits Syst. 07*, 2132–2141 (2016)
401. Skubiszewski, M. An exact hardware implementation of the Boltzmann machine. In *Proc. Fourth IEEE Symposium on Parallel and Distributed Processing* 107–110 (IEEE, 1992)
402. Peterson, C. & Söderberg, B. A new method for mapping optimization problems onto neural networks. *Int. J. Neural Syst. 01*, 3–22 (1989)
403. Söderberg, B. Optimization with neural networks. In *Scientific Applications of Neural Nets* (eds Clark, J. W., Lindenau, T. & Ristig, M. L.) 243–256 (Springer, 1999)
404. Okada, S., Ohzeki, M. & Taguchi, S. Efficient partition of integer optimization problems with one-hot encoding. *Sci. Rep. 9*, 13036 (2019)
405. Kanter, I. & Sompolinsky, H. Graph optimisation problems and the Potts glass. *J. Phys. A: Math. Gen. 20*, L673–L679 (1987)
406. Ceccarelli, F. et al. Recent advances and future perspectives of single-photon avalanche diodes for quantum photonics applications. *Adv. Quantum Technol. 4*, 2000102 (2021)
407. Stipčević, M., Wang, D. & Ursin, R. Characterization of a commercially available large area, high detection efficiency single-photon avalanche diode. *J. Lightwave Technol. 31*, 3591–3596 (2013)
408. Lu, X. et al. A 4-μm diameter SPAD using less-doped *n*-well guard ring in baseline 65-nm CMOS. *IEEE Trans. Electron Devices 67*, 2223–2225 (2020)
409. de Albuquerque, T. C. et al. Integration of SPAD in 28 nm FDSOI CMOS technology. In *2018 48th European Solid-State Device Research Conference (ESSDERC)* 82–85 (IEEE, 2018)
410. Morimoto, K. & Charbon, E. A scaling law for SPAD pixel miniaturization. *Sensors 21*, 3447 (2021)
411. Baltes, H. et al. *CMOS-MEMS: Advanced Micro and Nanosystems.* (John Wiley & Sons, 2008) https://doi.org/10.1002/9783527616718
412. Fedder, G. K., Howe, R. T., Liu, Tsu-JaeKing & Quevy, E. P. Technologies for cofabricating MEMS and electronics. *Proc. IEEE 96*, 306–322 (2008)
413. Qu, H. CMOS MEMS fabrication technologies and devices. *Micromachines 7*, 14 (2016)
414. Baltes, H., Brand, O., Hierlemann, A., Lange, D. & Hagleitner, C. CMOS MEMS - present and future. In *The Fifteenth IEEE International Conference on Micro Electro Mechanical Systems, 2002*, 459–466 https://doi.org/10.1109/MEMSYS.2002.984302. https://ieeexplore.ieee.org/document/984302 (2002)
415. Bugnacki, M. A micromachined thermal accelerometer for motion, inclination, and vibration measurement. *Sensors 18*, 98–104 (2001)
416. Kress, H.-J., Bantien, F., Marek, J. & Willmann, M. Silicon pressure sensor with integrated CMOS signal-conditioning circuit and compensation of temperature coefficient. *Sens. Actuators A Phys. 25*, 21–26 (1990)
417. Abadal, G. et al. Monolithic integration of MEMS resonators in a 0.35 μm CMOS technology for gravimetric sensor and radiofrequency applications. In *Integration Issues of Miniaturized Systems-MOMS, MOEMS, ICS and Electronic Components (SSI)*, 1–8. https://ieeexplore.ieee.org/document/5760511 (2008)

418. Montanyà, J., Valle, J., Barrachina, L. and Fernández, D. MEMS devices and sensors in standard CMOS processing. In *Solid-State Sensors, Actuators and Microsystems, Transducers Eurosensors XXVII*, 713–717. https://doi.org/10.1109/Transducers.2013.6626866. https://ieeexplore.ieee.org/document/6626866 (2013)
419. Fernández, D., Ricart, J. and Madrenas, J. Experiments on the release of CMOS-micromachined metal layers. *J. Sensors* (2010). https://doi.org/10.1155/2010/937301. https://www.hindawi.com/journals/js/2010/937301/
420. Michalik, P., Fernández, D., Wietstruck, M., Kaynak, M. & Madrenas, J. Experiments on MEMS integration in 0.25 μm CMOS process. *Sensors 18*, 2111 (2018)
421. Valle, J., Fernández, D., Madrenas, J. & Barrachina, L. Curvature of BEOL cantilevers in CMOS-MEMS processes. *J. Microelectromech. Systems 26*, 895–909 (2017)
422. Reitz, J. R., Milford, F. J. & Christy, R. W. *Foundations of Electromagnetic Theory* (Pearson/Addison-Wesley, *2009*). https://books.google.es/books?id=vNVDPgAACAAJ
423. Hung, Je-Min; Xue, Cheng-Xin; Kao, Hui-Yao; Huang, Yen-Hsiang; Chang, Fu-Chun; Huang, Sheng-Po; Liu, Ta-Wei; Jhang, Chuan-Jia; Su, Chin-I; Khwa, Win-San; Lo, Chung-Chuan; Liu, Ren-Shuo; Hsieh, Chih-Cheng; Tang, Kea-Tiong; Ho, Mon-Shu; Chou, Chung-Cheng; Chih, Yu-Der; Chang, Tsung-Yung Jonathan; Chang, Meng-Fan A four-megabit compute-in-memory macro with eight-bit precision based on CMOS and resistive random-access memory for AI edge devices. Nature Electronics (2021). https://doi.org/10.1038/s41928-021-00676-9
424. Chen, Ruibo; Liu, Hongxia; Song, Wenqiang; Du, Feibo; Zhang, Hao; Zhang, Jikai; Liu, Zhiwei Robust and Latch-Up-Immune LVTSCR Device with an Embedded PMOSFET for ESD Protection in a 28-nm CMOS Process. Discover Nano (2020). https://doi.org/10.1186/s11671-020-03437-3
425. Gavaskar, K.; Dhivya, R.; Dimple Dayana, R. Low Power CMOS Design of Phase Locked Loop for Fastest Frequency Acquisition at Various Nanometer Technologies. Wireless Personal Communications (2022). https://doi.org/10.1007/s11277-022-09654-6
426. Saadi, Hyem; Attari, Mokhtar; Escid, Hammoudi Noise Optimization of CMOS Front-End Amplifier for Embedded Biomedical Recording. Arabian Journal for Science and Engineering (2020). https://doi.org/10.1007/s13369-020-04347-3
427. Whitehead, William; Nelson, Zachary; Camsari, Kerem Y.; Theogarajan, Luke CMOS-compatible Ising and Potts annealing using single-photon avalanche diodes. Nature Electronics (2023). https://doi.org/10.1038/s41928-023-01065-0
428. Valle, J. J.; Sánchez-Chiva, J. M.; Fernández, D.; Madrenas, J. Design, fabrication, characterization and reliability study of CMOS-MEMS Lorentz-force magnetometers. Microsystems & Nanoengineering (2022). https://doi.org/10.1038/s41378-022-00423-w
429. D. Wu, K. Owen, B. Yu, and Y. Yi, 'Fabrication of a self aligned multi waveguide layer passive Si_3N_4/SiO_2 photonic integrated circuit for a 3-D optical phased array device', *Optical Materials Express*, **14**, 13 (2024)
430. Y. Yi, D. C. Wu, V. Kakdarvishi, B. Yu, Y. Zhuang, and A. Khalilian, 'Photonic Integrated Circuits for the Optical Phased Array', *Photonics*, **11**, 243 (2024)
431. D. Wu, V. Kakdarvishi, B. Yu, and Y. Yi, 'Photonic integrated circuit with multiple waveguide layers for broadband high-efficient 3-D OPA', *Opt. Lett.*, **48**, **Editor's Pick**, 968 (2023)
432. M. Ostrowski, P. Pignalosa and Y. Yi, 'Integrated disk micro resonator lasing using low-index ultra thin compound semiconductor thin films' *Nature Methods*, Accepted (2023)
433. D. Wu, B. Yu, and Y. Yi, 'Phase-Combining Unit for Aliasing Suppression in Optical Phased Array', *Optics Letters*, **47**, 1996 (2022)
434. A. Kazemian, P. Wang, Y. Zhuang, and Y. Yi, 'Machine learning based optimization of the aperiodic structure optical phase array antenna on silicon-on-insulator' *Optics Letters*, **46**, 801 (2021)

435. P. Wang, A. Kazemian, X. Zeng, Y. Zhuang, and Y. Yi, 'Optimization of aperiodic 3D optical phased arrays based on multi-layer Si_3N_4/SiO_2 platforms and machine learning', *Applied Optics*, **60**, 484 (2021)
436. D. Wu, Y. Yi and Y. Zhang, 'High efficiency end-fire 3-D optical phased array based on multi-layers Si_3N_4/SiO_2 platform', *Applied Optics*, **59**, 2489 (2020)
437. D. Wu, W. Guo, and Y. Yi, 'Compound period grating coupler for double beams generation and steering', *Applied Optics*, **58**, 361 (2019)

The manufacturer's authorised representative in the EU is Springer Nature Customer Service Centre GmbH, Europaplatz 3, 69115 Heidelberg, Germany. If you have any concerns regarding our products, please contact ProductSafety@springernature.com

Printed and bound by CPI Group (UK) Ltd, Croydon, CR0 4YY

26/03/2026

02078967-0007